CAMBRIDGE LIBRARY COLLECTION

Books of enduring scholarly value

Life Sciences

Until the nineteenth century, the various subjects now known as the life sciences were regarded either as arcane studies which had little impact on ordinary daily life, or as a genteel hobby for the leisured classes. The increasing academic rigour and systematisation brought to the study of botany, zoology and other disciplines, and their adoption in university curricula, are reflected in the books reissued in this series.

Memorials of Sir C.J.F. Bunbury

Sir Charles James Fox Bunbury (1809–86), the distinguished botanist and geologist, corresponded regularly with Lyell, Horner, Darwin and Hooker among others, and helped them in identifying botanical fossils. He was active in the scientific societies of his time, becoming a Fellow of the Royal Society in 1851. This nine-volume edition of his letters and diaries was published privately by his wife Frances Horner and her sister Katherine Lyell between 1890 and 1893. His copious journal and letters give an unparalleled view of the scientific and cultural society of Victorian England, and of the impact of Darwin's theories on his contemporaries. Volume 8 covers the years 1878–83. Family and social matters, and comments on the disturbed political state of Europe, and British military engagements in Africa and Afghanistan, predominate, although Bunbury also comments on plants he has seen and books he has read.

T0188058

Cambridge University Press has long been a pioneer in the reissuing of out-of-print titles from its own backlist, producing digital reprints of books that are still sought after by scholars and students but could not be reprinted economically using traditional technology. The Cambridge Library Collection extends this activity to a wider range of books which are still of importance to researchers and professionals, either for the source material they contain, or as landmarks in the history of their academic discipline.

Drawing from the world-renowned collections in the Cambridge University Library, and guided by the advice of experts in each subject area, Cambridge University Press is using state-of-the-art scanning machines in its own Printing House to capture the content of each book selected for inclusion. The files are processed to give a consistently clear, crisp image, and the books finished to the high quality standard for which the Press is recognised around the world. The latest print-on-demand technology ensures that the books will remain available indefinitely, and that orders for single or multiple copies can quickly be supplied.

The Cambridge Library Collection will bring back to life books of enduring scholarly value (including out-of-copyright works originally issued by other publishers) across a wide range of disciplines in the humanities and social sciences and in science and technology.

Memorials of
Sir C.J.F. Bunbury

VOLUME 8: LATER LIFE PART 4

EDITED BY
FRANCES HORNER BUNBURY
AND KATHARINE HORNER LYELL

CAMBRIDGE
UNIVERSITY PRESS

CAMBRIDGE UNIVERSITY PRESS

Cambridge, New York, Melbourne, Madrid, Cape Town,
Singapore, São Paolo, Delhi, Tokyo, Mexico City

Published in the United States of America by Cambridge University Press, New York

www.cambridge.org
Information on this title: www.cambridge.org/9781108041195

This edition first published 1892
This digitally printed version 2011

ISBN 978-1-108-04119-5 Paperback

MEMORIALS

OF

Sir C. J. F. Bunbury, Bart.

EDITED BY HIS WIFE.

THE SCIENTIFIC PARTS OF THE WORK REVISED BY
HER SISTER, MRS. LYELL.

———

LATER LIFE.

VOL. IV.

———

MILDENHALL:
PRINTED BY S. R. SIMPSON, MILL STREET.
———
MDCCCXCII.

1878.

JOURNAL.

The sad news of the death of Thornhills' son 1878. Compton.

Arthur came back with his sister Louisa.

Examined a Davallia from Java, and wrote a note on it.

The servants' dance—our young men danced all night.

The new year has opened upon us (I mean the nation), amidst threatening clouds and with a troubled and gloomy prospect. England is, thank God, as yet at peace, but the horrible war in the East seems likely to continue, and so long as it continues, it incessantly threatens to involve the other powers of Europe.—A war so savage and

1878. desperate, waged in such a ferocious spirit and with such inhumanity on both sides, and productive of such wide-spread misery, I hardly remember to have heard of in our time, (I speak of European wars); indeed I doubt whether there has been so horrible a war in this quarter of the world since that of Thirty Years. (Perhaps the earlier part of the war between France and Spain, in 1808 and 9, ought to be excepted). I cannot say that I feel much sympathy with either party—certainly not with either Government.

The Turks undoubtedly have shown a much greater fighting power than was expected ; the defence of Plevna was magnificent, and on various occasions they have shown that, in military matters at least, "the sick man's" condition is far from desperate. The struggle must have been very exhausting to the Russians, but still I think that their resources must be greater than those of their enemy.

Besides this formidable cloud in the East, we have the prospect of a Caffre war on our hands in South Africa, and a war with the Afreedees on the North-West frontier of India. At the same time, the distress in the colliery district of South Wales is terrible, almost as bad as a famine. May God help us!

———

Tuesday, January 8th.

Miss Bucke came.

The children's party—a very pleasant sight: Arthur presided.

News of the death of the King of Italy.

Clement went away—is going to embark for New Zealand.

The Barnardistons (3), Jos. Rowley and his sister, Miss Newton and her niece Ada, Colonel Ives, John Hanmer arrived: Lady Hoste and her son also dined with us.

A blowing day, but mild.

All the party except myself went to the Bury Ball.

A beautiful day.

Walked—returning, met Fanny, Lady Florence and Miss Newton, and strolled through the gardens with them.

Dancing again in the evening, and much merriment.

Another very fine day, though foggy in morning.

Our dance—a party of 17 from Stowlangtoft, also the Bevans, Lady Hoste and her son : the Nasons, about 44 in all—great merriment.

All our guests went away : Harry, George and Arthur alone remaining.

———

Dear Arthur and his sister Louisa left us : Harry also went to London.

George dined with the Wilsons.

My nephew George was gazetted yesterday to a commission in the 2nd battalion of the 9th regiment. He is very much rejoiced. I think he will make an excellent soldier. The corps to which he is appointed is stationed on the N. W. frontier of India.

Clement has gone off on a visit to Australia and New Zealand. Some time ago, rich Mr. Mackenzie invited him to go out as a sort of semi-tutor or *governor* with *his* son, an inexperienced young man, who is at Cambridge, and whom he wished to take that extensive trip, to give him some acquaintance with men and manners. Clement is to have all his expenses paid, and to have (I think) £200 besides ; and he hopes (and I hope) that the voyage will do his health good. He expects to be away hardly six months.

———

Fanny and George went to luncheon with Mrs, Abraham.

———

Mr. and Mrs. Livingstone arrived. Lady Hoste and her son, and the Percy Smiths dined with us.

Thursday, January 31st.

Harry returned to Cambridge and George went to London.

Mr. and Mrs. Livingstone went away, and we were (for a wonder) alone.

The aspect of public affairs is still very uncertain and uncomfortable. Since the last entry I made on the subject in this Journal, the course of the war has completely changed, and has become on the whole more alarming to those European powers, which (like ourselves) have not yet taken any part in it. The fall of Plevna was quickly followed by the capture of the Shipka pass, and of the whole Turkish army which guarded it; and upon this ensued what seems to be a total collapse of the Turkish military force. The whole state seems to be seized with a panic, and hopeless of defending itself; there appears to be no longer in any quarter, any prospect of effectual resistance. This sudden success of the Russian arms is embarrassing and, indeed, alarming to the Western powers. Russia has, indeed professedly granted an armistice, and negotiations for peace are said to be going on, but in the mean time the Russians are going on with their military operations, and nobody knows where they will stop, or whether they may not seize Constantinople and Gallipoli, without waiting for the negotiations. In

1878. fact nobody knows whether we shall be able to continue at peace, or be compelled to fight, not for the Turks, but for our own rights, and this being the case, I cannot blame our Government for making some preparations.

LETTER.

Barton, Bury St. Edmund's,
February 4th, 78.

My dear Katharine,

I thank you most heartily for your kind letter and good wishes, and also for the pretty letter-book which I am very glad to have. I may well be, and I hope I am, very grateful for having been allowed to reach my present age in such good health, and in the enjoyment of so many blessings, especially of so many good and kind friends, though at such times one cannot help looking back with a sad and longing regret to those whom one has lost. Yet it is not an unmixed sadness; I feel myself the richer for the delightful memories which I often recall of those whom I have admired and loved.

Fanny, I am thankful to say, is tolerably well. we took a walk with the dogs this morning. We have had a loss which grieves her much; the younger of the two goats, the English one, is dead; it was a very tame creature, very fond of her, and very pretty.

I heard glowing accounts of your ball; Arthur and his sisters, and Harry seemed to have enjoyed

it exceedingly. You were so kind to the young 1878.
MacMurdos. and I am sure your kindness is not
thrown away on them.

Our Mildenhall Vicar and Mrs. Livingstone spent
two days of last week, from Tuesday to Thursday
with us, and we got some of the neighbours to meet
them at dinner. We were much pleased with them.
Mr. Livingstone is a very well-informed accomplished man, very agreeable; he has travelled, and
was English Chaplain at Rome for (I think) two
years, and I am delighted by his love of books. I
hear too, that he has already made a very favourable impression at Mildenhall. His wife is lady-
like, very quiet and gentle. I hope I have again
been fortunate in my choice.

I have not time to enter upon the wide and
really awful subject of public affairs. As you say,
we are, indeed, passing through a critical period;
I hope God will mercifully preserve us from the
curse of war—though my young nephews are longing
for it.

I have gone through the two volumes of "The
Voyage of the Challenger" but with a great deal of
skipping, passing over all those matters which I
did not understand. I found much to interest me
in the descriptions of the islands they visited, and
the illustrations are most beautiful.

With much love to dear Rosamond,

Believe me ever,

Your affectionate brother,

Charles J. F. Bunbury.

(Aged 69).

JOURNAL.

We had the sad news (in a letter from Lady Grey to Fanny) of Sir Frederick Grey's dangerous illness. The doctors seem to think there is no hope. I am much grieved. Our acquaintance with them began at Madeira, in the beginning of '54, and has gradually gone on increasing till we have become very intimate, and in the same proportion our liking, esteem and regard have increased. The husband and wife are so united, so devoted to each other, so happy in each other, in their home, in their circumstances and their pursuits—it is very, very sad.

———

Fanny Mallet lent me to read, two very long letters written to her by May Frere, from the Cape. They are very interesting and wonderfully clever. They give me a high opinion of Miss Frere's ability; I am much struck, not only with the brilliancy and vigour of expression, but with the sagacity and shrewdness of her observations. By her account the state of affairs at the Cape seems to be very uncomfortable, especially for the Governor, who finds scarcely anybody able to understand the real state of affairs, or to co-operate effectually with him in securing the frontier districts against the enemy.

———

February 12th.　1878.

We have had a very agreeable party in the house since the 5th :—Mr. and Mrs. John Martineau, Lady Susan Milbank, Sir Francis Doyle and Miss Doyle, Louis and Fanny Mallet, Minnie Powys, Mrs. Wilson and Agnes, Mr. Egerton Hubbard, Edward, and for part of the time my nephew George.

LETTER.

Barton,
February 13th, 1878.

My Dear Susan,

I thank you very heartily for your good wishes on my birthday.

We are just now quite alone, which is very unusual, but not at all disagreeable to us, as we have always plenty to do, and it is very good for Fanny to have an occasional respite from the cares of society. You will hear from her about the very agreeable company we had in the house last week.

Sir Francis Doyle (who was for the first time here), is a clever and very well-read man ; was for some time Professor of Poetry at Oxford, and has a very extensive acquaintance, both with books and men, and has a wonderful store of anecdotes. Fanny was delighted with him, but my deafness (which is generally a great plague to me in society), prevented me from profiting thoroughly by his con-versation.

1878.   I am reading "Democracy in Europe," by Sir
Erskine May ; I was so much pleased with his
"Constitutional History of England" (a sequel to
Hallam, beginning with the accession of George the
Third), that I expected to be much interested, but
I am disappointed with the first volume.   It is
pleasantly written, but not impressive : and as far
as I have hitherto read, found little that is new
to me.

There is a new book by Mr. Lecky, "A History
of England in the Eighteenth Century," which
promises to be interesting, but I cannot begin it just
yet.

Of Kingsley's "Life," I am told, 10,000 copies
have been sold, or at least ordered.   I am heartily
glad that my dear friend Mrs. Kingsley has been
gratified by such a success of her work, and that she
has contributed—not only to extend her husband's
fame—but to keep alive so effectually his usefulness
and influence for good.

I am amused by hearing of your reading "Ouida,"
I should not think you would find in her much
either to please or instruct ; indeed I have not read
any of her books, but I have dipped into several.

Has the Pope's death made much sensation in
Italy ?   Not very much I should imagine ; poor old
Pio Nono, I take it, had outlived his importance :
and I rather suppose it does not matter much who
may be his successor, unless it should happen to be
a really liberal-minded man, which is not likely.

I am much interested by your remarks on books.
Washington Irving's "Life of Washington" I never

read, but think of doing so, as it is sure to be 1878.
interesting under his treatment, and I am not at all
well read in the subject.

Of the Etruscan language I know nothing, but
have a sort of vague recollection that Niebuhr
theorized on the derivation of the Etruscans from a
Teutonic stock.

Pray give much love from me to dear Joanna,
Leonora, Annie and Dora, and thank Joanna very
much for her letter to me, and believe me ever,

Your affectionate brother,

CHARLES J. F. BUNBURY.

JOURNAL.

February 15th.

Since I first was able to understand something of
politics, and to take an interest in them, I can hardly
remember a crisis (in our foreign relations) so
strange, so perplexing, so alarming and so disagreeable as the present. The Turks losing courage
altogether after the capture of Plevna and of their
Balkan defences, seemed to have given up the
struggle in despair, and surrendered at discretion,
placing themselves entirely at the mercy of Russia.
The Russians have not indeed, actually entered
Constantinople, but it seems to be in their power to
do so whenever they choose, for the *Lines* which
defended it have been given up to them. Our fleet
has been ordered up, and in spite of a protest from
the Turks, has advanced as far as the Princes
Islands, a very few miles from Constantinople.

1878.   We seem to be on the very brink of a war, and multitudes in this countr yare clamouring for war, but with whom ? or why ? Our " ancient allies " the Turks, have placed themselves at the disposal of Russia, so that if we go to war with the one, we must fight against the others also.

We seem likely to be driven into a dangerous war without any intelligible cause or object. Our fathers, in their great war against revolutionary France, had at any rate a clear and intelligible principle to fight for, whether right or wrong; they fought for monarchy and aristocracy, against the democratic propagandism of France. But it is wretched to be plunged into the midst of the horrors and dangers of war, without knowing what we have to fight for. Happily peace does not yet appear quite hopeless, and there is nothing for which we ought more earnestly to pray.

February 21st.

The Conclave has been remarkably prompt in its operations. A new Pope is already elected— Cardinal Pecci, who is to bear the name of Leo the Thirteenth.

March 1st.

The weather during almost the whole of February has been very remarkably fine and dry. I hardly remember a February with so large a proportion of bright and sunny days, so little of fog and gloom ;

and I should think the rainfall must have been far 1878.
below the average.

News of the signature (at last) of the Treaty of
Peace between Russia and Turkey. There, at last
is something gained, and I trust that Russia will be
wise enough to give us no excuse for the wicked
folly of going to war.

In one of these last three days, I have read Sir
Garnet Wolseley's article (in the *Nineteenth Century*,)
on " England as a Military Power in '54 and in '78."
Very clear, full, and interesting; and so far satisfactory, that it shows that we are as well prepared
for war as we are ever likely to be without a system
of conscription, from which system *Libera nos
domine !*

Dear Kate Hoare, with her husband and her
elder little boy, arrived.

I have omitted to mention in its place, that, while
we were alone, Fanny read to me several sermons,
newly published, by Canon Farrar, against the
doctrine of Eternal Punishment. The style is too
florid and rhetorical for my taste: but in every other
respect I admire and approve them heartily.

She has also read to me several sermons by Mr.

1878. Haweis, which are in a very peculiar style; often very odd, sometimes strangely familiar, even grotesque and laughable, often very powerful and really eloquent.

March 11th.

Fanny read to me a very fine sermon of Kingsley, in his newly-published posthumous volume of sermons (edited by his son-in-law, Mr. Harrison):—the same which we heard him preach in Westminster Abbey on the 27th of September, 1874. It is number 19 in the posthumous volume, entitled "Good Days."

March 12th.

Fanny read to me one of Kingsley's sermons, included in the posthumous volume which I mentioned yesterday:—its title is "Human Soot; and its purpose is to urge the duty of rescuing and helping the outcast of society, and especially outcast and neglected children. It is a truly noble sermon.

March 14th.

Good hope now, I trust, that peace will not be broken as far as we are concerned. The peace between Russia and Turkey has been signed; the Great powers appear to have agreed on a Congress to be held at Berlin, to settle definitively the terms of an agreement on the Eastern Question; and thus it may be hoped that time will be allowed for all tempers to cool down. Though Russia— 1878. with a strange and certainly unsatisfactory secrecy —still avoids publishing the actual terms of the treaty, there is reason to believe that none of them are such as we need resent ; and in spite of the very unreasonable philo-Turkish zeal which has somehow been excited in England, I am in hopes that we shall avoid war.

March 17th.

My rejoicing was sadly premature. The prospect of peace is again clouded over.

Kate Hoare and her husband left us on the 15th ; Lady Muriel and Mr. Courtenay Boyle yesterday. We have had a nine days' spell of very agreeable society :—besides these four, Mr. Bowyer and John Hervey (for part of the time), and our neighbours Mrs. Wilson and her daughter Agnes, Mr. Greene and Lady Hoste, Mr. and Mrs. James, as guests at dinner.

Kate's heart is still, of course, very full of Sarah, and I had some most interesting and touching conversation with her on the subject : but she has in general recovered much of her former cheerfulness.

My opinion of Lady Muriel Boyle and her husband is as favourable as last year. Lady Muriel I like particularly.

Mr. Courtenay Boyle, a good, earnest, thoughtful man, cultivated and agreeable.

There are not many women whom I like better, or esteem more, than Mrs. Maitland Wilson.

1878.   March 25th.

The dreadful disaster of the Eurydice, a training ship of the Royal Navy, capsized in a sudden snowstorm, off Ventnor, with more than 300 men on board, of whom only two survived.   It is indeed a most truly awful calamity.

March 29th

A heavy fall of snow, continuing till near 2 p.m. with a violent wind, but no great degree of frost : much of the snow was not melted by night time. In the latter part of the afternoon, that part of the lawn from which the snow had melted, was covered by an extraordinary number of Starlings, which remained there a long time, feeding eagerly : they quite blackened the ground.

Very bad news : Lord Derby has resigned.   I much fear this is an omen of war—an omen that the violent war party is gaining a decisive ascendancy in the Ministry.   Everything looks dark and threatening.

March 31st.

This day, 37 years ago, I first saw her who is now my wife.   How good God has been to me !

LETTER.

Barton, Bury St. Edmund's,
April 2nd, '78.

My Dear Katharine,

I am exceedingly glad to hear that you

and Arthur have arrived safely in town, and that he did not seem the worse for the journey. I trust he will soon get all right. I trust you have not thought me unfeeling for not writing sooner to express my sympathy with your distress and anxiety about him ; I hope you know me better. I know I ought to have written, and I was often thinking of it, but too often—"the things that I would I do not."

During this quiet time, while we have been quite alone, without any particular business, I have often wondered how the days have flown away. I never find them long enough for half of what I want to do.

I have been writing a good deal of my botanical notes, and am beginning to have some ideas as to putting them into shape.

We are reading Lecky's new book together (that is, I read it aloud in the evenings, and re-read it to myself) : it is very interesting, and like all his writings, has a peculiarly agreeable style. I am also reading Green's "History of the English People," which has been republished in a better form and type, and is both instructive and interesting, though I cannot say that it has the merit of an easy style.

What a burst of winter we have had! Melancholy for the poor little birds, which must have been thinking through most part of March, that Spring had come in earnest. To-day the wind has changed, and it is milder, but stormy.

Fanny has gone over to Mildenhall to see poor

1878. good old Mrs. Marr, who, I fancy, is not long for this world.

Fanny is just now returned safe and well from Mildenhall, just in time to escape a heavy storm of rain.

I wish you could see our *houses* : they are now in great beauty, especially the cool greenhouse, with its azaleas, camellias, passion flowers and Australian plants.

With my best love to Rosamond and Arthur,

Believe me,

Your affectionate brother,

CHARLES J. F. BUNBURY.

JOURNAL.

April 3rd.

Read Lord Salisbury's Circular to the diplomatic agents of this country. A very important document, extremely able, very clear and very vigorous. It seems calculated to make a considerable sensation. It has thrown a new light on that perplexed Eastern question, and made me understand it all far better than I did before. It clearly shows what I certainly never understood before, that the question is not one merely of diplomatic formulas and subtleties, but that our Ministry object to the Treaty of San Stefano as a whole, and to all the principal provisions of it. Their objections are very clearly and forcibly stated by Lord Salisbury.

I am afraid that Lord Salisbury's Circular is hardly likely to have a conciliatory effect upon the

Russians; but it is very satisfactory, after so much 1878. diplomatic verbiage and subtlety, so much that is vague and indistinct, and would be mysterious, to meet with a document which speaks out plainly and clearly, says what it means, and goes direct to the points. If we must have war, now we shall at least know what we are fighting for. Yet I think even our Ministers can hardly expect Russia to yield up, on our demand alone, all the fruits of a war which has cost them such enormous sacrifices. Are we to wage war with the object of replacing Turkey in the position she was in two years ago? What a chimera!

———

April 10th.

Read (this day and the day before) all the most important speeches in the great debate, in both houses. Lord Beaconsfield's, ingenious and plausible; Lord Derby's, clear, open, frank, manly, and (to me) almost convincing; Lord Salisbury's, sarcastic, bitter, and mischievous; Sir Stafford Northcote's, fair, temperate, conciliatory, and eminently gentleman-like; Gladstone's eloquent, brilliant, impetuous, not giving one the impression of a *safe* leader.

———

April 11th.

Gortschakoff's reply to Lord Salisbury's Circular appears to me temperate and reasonable enough. and I cannot see why we might not go into the Congress on the basis afforded by it.

1878.   April 12th.

Grieved and shocked by hearing of the almost sudden death of Margaret Richmond, my cousin Aberdare's eldest daughter. I am indeed very slightly acquainted with her husband, and have seen her but little since her marriage; but she was an excellent and accomplished person, and seemed very happy, and her death is a grievous affliction to her poor father and to my cousin Norah.

===

LETTER.

Barton,
April 12th, '78.

My Dear Edward,

It is possible you may not have heard of the sad misfortune which has befallen the Aberdares. His eldest daughter, Margaret Richmond, died two days ago, of inflammation of the lungs, after a very short illness. We have had the news from Katharine Lyell, and are much shocked and grieved by it.

What do you think of the prospects of war or peace? We hear that you are a zealous philo-Turk. I am not; I do not indeed particularly admire the Russians, but I am strongly against war; I see hitherto no necessity for it, and I cannot perceive what we are to fight *for*—for what cause or for what object? Supposing we are ever so victorious, what then? what are we to do next? Surely we are not going to fight for the chimerical object of restoring

the Turkish Empire to what it was three years ago ?
Lord Salisbury's Circular was not only very able,
but had the great merit of being extremely clear and
straightforward, and dissipating the diplomatic
mists, but now that Gortschakoff has answered it in
a temperate and conciliatory tone, I do not see why
we might not go into a Congress with some fair
hope of coming to an agreement.

With much love from Fanny, believe me,

Ever your affectionate brother,

CHARLES J. F. BUNBURY.

JOURNAL.

April 13th.

Edward writes to me from London, of the change
which " has somehow or other come over the state
" of public feeling within the last few days; nobody
" seems to know why, but everybody seems now
" to think there will not be war, whereas, ten days
" ago everybody thought there would be."

" Lord Salisbury's dispatch," (he adds), " has
" produced a great effect on the Continent, and
" seems to have called forth an amount of sympathy
" and good feeling towards England, that cannot be
" without some effect even on such a power as
" Russia. But no doubt there is still much risk, and
" it will be long before things settle down into any-
" thing like security and tranquillity."

1878.   April 20th.

Very agreeable company since the 6th. Dear
Minnie arrived on that day, and has been with us
ever since. John Herbert came on the 8th, and has
left us on the 20th. Dear Sarah and Albert
Seymour, with their darling little boy, came on the
8th and left us on the 18th. Lady Mary Egerton
and her daughter Charlotte were with us from the
8th to the 12th; Mr. and Mrs. Clements Markham
and Captain Markham from the 9th to the 13th;
Mr. and Mrs. Hutchings from 15th to the present
time. We have also repeatedly had several of our
neighbours to meet them at dinner, particularly
the Wilson and Hoste families. It has been a very
agreeable time.

––––––

April 23rd.

I have the great satisfaction of hearing a very
good report of Mr. Livingstone, the new Vicar of
Mildenhall (whom I nominated last year, both from
Mr. Lott and through Scott), from several of the
principal people of Mildenhall, and from Dissenters
as well as Churchmen. All that I hear of him is in
his favour. He was in a difficult position at first,
succeeding to so active and energetic a clergyman
as Mr. Robeson, and coming, too, when the Church
in Mildenhall had been much disheartened and
disconcerted by Mr. Robeson's unexpected resignation. But Mr. Livingstone seems to be gradually
and quietly winning his way and gaining golden
opinions. He seems to be active—but not bustling

or ostentatious—gentle, conciliatory, charitable ; on 1878. good terms with Dissenters as well as Churchmen. It will be a great mercy if my *second* choice of a clergyman for Mildenhall turns out as well as the first.

_____

April 25th.

Mr. Harry Jones and Mrs. Jones came to luncheon with us, and we showed them our green-houses.   Mr. Jones is an excellent man, working immensely hard and doing a world of good in the east of London (where his living is), and at the same time a remarkably active-minded energetic man with great powers of observation, and a mind awake and alert on a variety of subjects.   He is a friend of the Dean of Westminster, sympathized with Kingsley, and admires the writings of Lecky.

_____

April 26th.

Beautiful spring weather lately.   The spring is in its full beauty, of flowers and leaves and birds ; it always gives me fresh delight every year.   I never grow tired of it.   Thank God I have not yet lived to know the days when I must say—" There is no pleasure in them."

That *may* yet be my lot.   If so, may God grant me patience and resignation, such as my dear brother had in his terrible trial.

_____

April 29th.

Heard the Cuckoo this morning, very clear and

1878. seemingly near; the first time this year, and, I understand that no one else hereabouts has heard it before. The swallow kind also are late this year; yesterday was the first time that I have seen any, and I think from Scott's account, that few have yet been seen by others. Those I saw were martins.

————

May 4th.

We received the very sad news of the death of Sir Frederick Grey; not unexpected indeed. He is a great loss. He was an excellently good and very able man, and a very agreeable one, with very lively, frank manners, and a variety of knowledge; in fact, the activity and versatility of his intellect, as well as its clearness, were remarkable; he seemed to be deeply versed in many of the physical sciences; and showed a peculiar readiness and ingenuity in all the practical operations connected with them. I felt a painful surprise when I first heard of his illness, for he had always seemed to me the very picture of health, vigour and activity, both bodily and mental.

————

May 5th.

Fanny had a letter from Clement, from Sydney— very pleasant in itself, and especially so as assuring us of his safety; for we have been a long time without direct news of him, and there had been disquieting rumours about the ship. And not without reason. He seems to have had a most prosperous voyage, and enjoyed it extremely, till

Friday, the 15th March, when they were within a   1878.
few hours of Sydney.   On that day, about 11 in the
morning, as they were running along very near the
shore (the weather being fine but rather hazy)—he
and many other passengers were writing letters in
the saloon, when suddenly the screw stopped ; the
next minute they felt a tremendous blow, then
another, together with a *grinding* sound, which he
says he shall never forget.   Knowing at once that
the ship had struck, he rushed on deck, and found
that they were within a little distance of a wall of
rock, 300 feet high, and extending several hundred
yards to the right and left.   It is a great mercy that
the weather was very fine and calm ; their safety
was owing (under God) to this and to the water-
tight compartments in which the vessel was
constructed ; two of these gave way and filled, but
the rest kept her afloat after she had backed off the
rock.   The passengers were safely landed in a little
cove, and the next day, they and their luggage were
sent in another steamer to Sydney, from whence he
wrote on the 17th March.

He and his friend Mackenzie were to go on from
thence in about ten days on their way to New
Zealand.   Altogether it was a most merciful escape.

————

May 9th.

After a long period of beautiful and delightful
weather, we have had almost a deluge.   The rain
began in the morning of the 7th.   It poured almost
continually during the whole of that day and the 8th ;

1878. not that the rain was violent, but I have seldom (of
late years at least) seen it continue steadily for so
long. In the 24 hours ending on the morning of
the 8th, there fell 1·46 inches of rain ; in the following 24 hours, 0·73in.

May 10th.

Up to London by the mid-day Ipswich train ;
had a good journey and arrived safe—Thank God.
The weather had improved, and the country is in
its loveliest spring dress ; the meadows in all their
brilliancy, the trees fully clothed with their young
leaves, and the fruit trees in their richest beauty of
blossom. It seems a pity to leave the country in
such a season.

May 14th.

We dined with the MacMurdo's at Rose Bank.
I think I have elsewhere noted the beauty of the
Macmurdos' house, and its remarkable wealth in the
most lovely flowers.

After dinner, a Mr. Pidgeon exhibited one of the
strange new inventions, the *Phonograph*.

May 16th.

Dear Cissy and Emily arrived, having come to-
day all the way from Paris ; Cissy dined with us.

It is a great pleasure to see our dear sister again
among us, after her long absence from England
(from August, '76 to May '78). It is unlucky that

she and her delicate daughter should have returned 1878.
in such rough, boisterous and chilly weather.

Went to the Botanic Society's Gardens in the
Regents Park, to buy tickets for the flower-show on
the 22nd; then walked through the great conservatory and about the gardens, which are very
pretty and pleasant. I have often noticed the
very skilful arrangement of those grounds, by which
the space is made to appear far larger than it
really is, and a thoroughly rural appearance is
given to a spot which is actually in London. And
now, in this beautiful spring weather, these gardens
looked especially delightful, with the profuse
blossoming of the hawthorns, laburnums, lilacs
and other trees and shrubs, the delicate tints of
the young foliage, the brilliant green of the grass,
and all lighted up by a clear sunshine.

The great conservatory contains a handsome
collection of fine tropical and sub-tropical plants:
Palms, Seaforthia elegans, especially fine: Tree-
Ferns, especially Dicksonias, herbaceous Ferns,
Aroids and others: many very beautiful, but I did
not observe any new to me.

Thence to the Zoological Gardens, which also
are in great beauty, and in a state of great
animation—the birds especially being much excited
by the season and the fine weather. Saw the Birds
of Paradise—new acquisitions and great rarities,
obtained from Paris. They appear to be healthy

1878. and lively, and are in full and beautiful plumage, exactly as Wallace describes them. I did not hear them utter any note. There are two of them, both alike in plumage. I rather think, from what I remember of Wallace's descriptions, that they belong to the smaller species, not to the largest. As both of them have the great development of delicate floating plumes at their sides, I suppose that both are males.

I find by the *Illustrated London News* (in which there is a good wood-cut of these birds), that they *are* both males, and that they do belong to the smaller species, the Paradisea *Papuana*.

In the new lion house I observed that the curious woolly-haired Cheetah which I noticed last year, is now distinguished as a peculiar species, under the name of *Felis lanea*, and marked as a native of S. Africa,—the true Cheetah is marked as native of India. It (the F. lanea), is remarkable not only for its curiously *crisped* or woolly hair, but also for the paleness and indistinctness of its spots.

May 21st.

We went yesterday evening (Admiral and Lady Sotheby, Minnie and my niece Emmie with us), to the Court Theatre, to see "Olivia," a new play founded on The Vicar of Wakefield, but deviating considerably from the novel. The humorous parts of the original, which are so delightful, are all omitted in the play. There is no visit of Moses or of the Vicar to the fair: no "gross of green spectacles:" no Mr. Jenkinson with his "Cosmogony 1878. or creation of the world:" no Lady Carolina Wilhelmina Amelia Skeggs. But the delicate simplicity and pathos is brought out with great effect, and the moral tone, the moral feeling of the whole is excellent.

It is a very pretty play, a sort of sentimental comedy, and we were delighted with the acting, especially that of Miss Ellen Terry, who was Olivia. Her acting, especially in the scene where she takes leave of her sister and young brothers, they supposing that she is only going to stay with a lady in London, while she has secretly agreed to elope with Thornhill: and the last scene, the reconciliation with her mother, and ultimately with her wicked husband and would-be seducer is quite exquisite—most natural and un-affected, and almost too pathetic. The scene with Thornhill in which he at last, after great persuasion, induces her to elope with him, is also beautiful acting. She is extremely handsome but appeared to me almost too dignified for such a giddy girl as Goldsmith's Olivia. Sophia also was well and agreeably acted, and Mr. Herman Vezin was a good Vicar.

———

May 21st.

Dear Isabel Hervey came to stay with us.

Minnie, Augusta Freeman, Cissy and Emmy, dined with us. I had not seen Augusta for some years; she is looking very well, and I do not see that she is at all altered, either in looks or in her

1878. peculiarly frank, hearty, straight-forward style. She is an excellent person.

———

We went (Minnie and Augusta, Cissy and Emmie with us), to see the flower-show at the Botanical Society's Gardens. Very much pleased. The display of flowers and ferns in the great exhibition tent was wonderfully beautiful, and arranged so as to produce the most delightful effect. The variety astonishing, and the specimens most of them superb. Noticed particularly, magnificent specimens of Anthurium Scherzerianum, with multitudes of its scarlet spathes: some other Aroids with superb leaves; Tillandsia Zebrina, with dark striped leaves and with a very curious and showy flowering spike, composed of closely imbricated distichous bracts of a brilliant scarlet colour. Sarracenia purpurea, in larger and more vigorous clumps, and with greater abundance of flowers than I remember to have seen it before. A splendid display of Orchids : magnificent Cattleyas, Odonto-glossums, Cypripediums, a brilliant crimson Mas-devallia and many more.

Very fine Ferns, among them two new Adiantums palmatum (very beautiful and distinctly marked), and cyclosorum and some really superb Tree-Ferns.

———

We went down on the 25th to the William Napiers at Sandhurst, and spent a day-and-a-half

with them, returning in the middle of the day on the
27th. The enjoyment of our visit was a good deal
marred by the very bad weather, but I managed
to have a pleasant walk with William between two
storms. The family party were William and Emily,
their daughters Susan, Georgina, and Emily; their
son Charles and their son-in-law Cecil Bunbury;
and as it was a Sunday when we were there, there
was no company. William is still the same pleasant
fellow and thorough gentleman, and excellent man
that he has always been. Susy (Cecil's wife) is
charming ; the other two girls very nice.

The country about Sandhurst has a peculiar
character of its own,—very interesting and strongly
characterized with much beauty of a special kind.
It is that country of sand and heath and fir trees,
hilly, yet without great hills,—which Kingsley often
described, and in which he delighted. To me also
it is a very pleasant and attractive country ; to
be sure, I have seen it mostly in summer. Just now
it is in great beauty, with the long and delicate
young shoots on the fir trees, the broom and furze in
full blossom, and the brilliant rhododendrons and
azaleas (which have been planted in abundance
among the fir trees and ferns, and grow as if they
were natives of the soil), fast coming into flower.

———

May 29th.

I see in *The Times* the announcement that Earl
Russell died last night. He was in his 87th year,
having been born on the 18th of August, 1792. His

1878. was a remarkable and honourable career, as well
as a long one.   He was not (in my opinion) a great
statesman, but he was a very honest and a very
useful one ;—eminently respectable ; he made the
very utmost and the best use of the gifts he had,
which were, I take it, in themselves by no means
extraordinary.   It is remarkable how completely he
seems to have outlived his *political* generation—outlived all who were conspicuous, either as his
comrades or his opponents in his early career.   For
many years past he has appeared to stand quite
alone.

_____

May 30th.

The 34th Anniversary of our marriage.   I humbly
thank the Almighty for having granted me so much
happiness as has been produced to me by this
union.

We went to the South Kensington Museum, and
looked at the collection of pictures lent by Lord
Spencer ; but we were so late, we had time for little
more than a glance at them.   There are many
beautiful portraits of the Spencer family, particularly by *Reynolds*.

_____

May 31st.

Dear Rose Kingsley came to luncheon ; it was a
very great pleasure to see her again, though but for
a short time, after being so long without meeting.
She sets off almost immediately for Aix-les-Bains in
Savoy, the waters of which have been recommended

for rheumatism. She is looking well, however.
We had a luncheon party to meet her : — May
Egerton, Mrs. Henry Grenfell, Minnie Napier,
Cissy and Emmie, besides Isabel Hervey, who is
staying with us.

Later in the afternoon, we, with Minnie, went
to the South Kensington Museum (the Exhibition
buildings), and saw Miss North's flower paintings;
but we had such an enormous walk to reach them,
through the endless shabby galleries of that most
ill-arranged building, that we were almost too much
tired to enjoy them thoroughly. Miss North's
admirable paintings I saw last year, and noted them
in my Journal of that time ; and I now admired
them as much as before.

June 1st.

Our dinner party; the Charles Hoares, Mr.
Courtenay and Lady Muriel Boyle, Philip and
Pamela Miles, the MacMurdos (including Mimi),
Mr. and Mrs. Walrond, Mr. and Mrs. Rycroft,
Admiral Spencer, Mr. Medlycott, Edward, Cissy,
Isabel, besides our two selves. A pleasant party.
I sat between Lady Muriel and Kate, both charm-
ing. Several others came in after dinner.

June 3rd.

We went to see the Water-colour Institute
exhibition, and were very much pleased with it.
There are very many delightful pictures. I noted

1878. particularly :—by *Edmund Warren*, several charming pieces of woodland scenery, with venerable
beech trees, mossy banks, forest glades, and delicate
peeps of distance, exquisitely done; by *Philp*,
beautiful views on the Cornish coast, with the
granite cliffs and the sea shown to perfection; by
H. G. Hine, an old Chalk Pit near Eastbourne, very
true to nature; "Peat Carriers Resting," North
Wales, and "Gathering Fern," by *J. H. Mole*,
and some charming little flower pieces by *Mrs.
Duffield*.

Two extraordinary catastrophes within these very
few days. Last Friday, there was the incomprehensible collision between two great German
ironclads, off Folkestone, with the sinking of one of
them, and a shocking loss of life; and this morning
the papers announce an attempt (the second within
a little while) and almost a successful one, to
assassinate the Emperor of Germany. It is horrible
—the assassination of a man of more than eighty
years of age, blameless in all his private relations,
and whom one can, at worst, only consider as mistaken in his political measures.

———

June 4th.

We went to the old Water-colour Gallery, in
Pall Mall East : there are many excellent pictures,
but on the whole we did not like it as well as the
other, at least as to the landscapes. There
however more variety, for there are many more
figure-subjects in this, and two by *Sir John*

Gilbert pleased me much; one called "For the 1878.
King," a fight in the great civil war—seemingly a
struggle for a convoy of waggons; the other—
"Travelling in the Middle Ages," apparently a
party of banditti, watching a large party of travellers
who are probably strong enough to make them
hesitate as to attacking.

--- --- ---

June 10th.

It is a great satisfaction to know that everything
is at last settled for the meeting of the Congress,
and that the prospect of peace is extremely fair.
Not that we are yet really "out of the wood," it
is still possible that, in the Congress itself, the
demands and pretensions of Russia may be found
entirely incompatible with the settled views of the
other Powers as to what is safe and right; but a
great deal is gained by bringing about a meeting
of representatives of all the great Powers for a full
discussion. It is also a great comfort to know
that the French Government is entirely with us;
nothing can be better than M. Waddington's speech
on the subject.

June 10th.

Our dinner party; Mr. Mills, Leopold and
Lady Mary Powys, Minnie Powys, Mr. and Mrs.
Cyril Graham, Isabel Hervey, Admiral Spencer,
Edward. After dinner, Mrs. Rycroft, Mrs. Rowley,
Miss Berners (these three ladies are sisters), Cissy
and Emmie—a pleasant party.

--- --- ---

                                        June 12th.

Fanny received, last night, a very pleasant and
interesting letter from Clement, written from *Napier*,
New Zealand, on Easter Sunday. He writes in
excellent spirits, seems delighted with New Zealand
and writes in raptures about an expedition which he
and others had made, to a region of *geysers* and
boiling lakes and fountains, about 200 miles from
Auckland. He describes vividly the extraordinary
natural phenomena of this region—the vast fountains
and basins and lakes of boiling water, the beautiful
terraces of silica and other minerals deposited from
them.

He was delighted also with the grand and luxuriant vegetation of the woods, which seems to be
almost tropical in its vigour and exuberance of
growth (though not in its variety) with the profusion
and beauty of the Ferns, and especially with the
magnificent Tree Ferns. He was struck with the
entire absence of all " beasts " (mammalia), and
the rarity of birds, except pheasants, which were
introduced in recent times by the English, and have
become so numerous as to be (he says) absolute
pests.

Received a short but pleasant letter from Rose
Kingsley, who has arrived safely at Aix-les-Bains,
and seems to be very happy; she is comfortably
established in the same hotel with Lord Charles
Hervey and his family, and very much enjoys their
society.

———

                                        June 13th.

I see in the newspapers the deaths of the ex-king

of Hanover, and of Bryant the American poet: the 1878.
former at the age of 59, the latter of 84.

I have read some of Bryant's poems with great
pleasure.

June 14th.

The good news that my nephew Harry has
succeeded in passing his " Previous Examination "
(familarly, *Little Go)*, at Cambridge. I am very
glad. It is a much more serious and important
examination than it was in my time.

June 15th.

Fanny had a letter from Mrs. Horton, who is now
at Baveno, telling us the good news that her
daughter Freda Broke is engaged to be married
to Sir Lambton Lorraine, a distinguished naval
officer. Of this also I am very glad, as the engagement seems to give great satisfaction to the family,
and I hope the marriage may be happy for Freda.
She is a remarkably nice girl.

June 18th.

Our dinner party; Lord and Lady Hanmer, Sir
Edward and Lady Blackett, Minnie Powys, the
Louis Mallets, Lady Head, Mr. and Mrs. Gambier
Parry, Mr. Egerton Hubbard, Mr, William Gurdon,
Mr. de Grey, Miss Doyle and her brother, Minnie
Napier, Isabel Hervey, Edward. After dinner
Lady Mary Egerton and her daughter Emily,

1878. Susan MacMurdo and her two daughters, Mr. Cyril
Graham, the Leonard Lyells, and many more.

A visit from Norah Aberdare—very agreeable, as
she always is. She is one of the most admirable
women I know. What she told me about the distress in the colliery district of South Wales, in
the midst of which they live,—the efforts required to
supply the people with necessary food,—the continued good behaviour of the poor people under
these severe privations, their gentleness and gratitude to those who helped them, and the entire
absence of crime—even of pilfering—was very interesting.

Willoughby and Mrs. Burrell, Minnie Powys,
and Minnie Napier came to luncheon. Mrs. Willoughby Burrell (who is the daughter of a physician
at Dublin), is an uncommonly clever, well-informed,
and agreeable person.

We dined with the Locke Kings; met nobody
whom I knew, except Lord Kinnaird. I sat by
Lady Kay Shuttleworth, who is pretty and agreeable.

We dined with Minnie; met John Herbert, old
Mr. and Mrs. Drummond, Mr. and Mrs. Marsham,

Miss Harriet Moore, and others. Mr. Marsham has been a great traveller, and is a very well informed and agreeable man.

- - - - -

We, with Minnie, went to the Savoy, and heard a really beautiful sermon from Mr. White, on the text, "Consider the lilies of the field." It sets forth in a peculiarly interesting and engaging manner the moral and spiritual lessons—the instruction as to the conduct of our lives and the regulation of our minds and thoughts—to be derived from the beauties of nature.

While we were in church, in the middle of the sermon, a violent thunderstorm began ; it seemed to burst directly over our heads ; the first two or three claps were so abrupt and sudden, they sounded more like the discharge of canon than ordinary thunder. The storm continued pretty heavy till near 3 p.m. The morning had been warm, but with a remarkable haze.

- - - - -

We visited the Zoological Gardens. The weather was beautiful, and the animals in a very lively condition. We met with one of the keepers—*Church* by name—who is known to Fanny, and he showed *off* several of the birds for us, enabling us to see them nearer and to more advantage than we could otherwise.

The Secretary bird—a very fine specimen—very

1878. tame with the keeper, who excited it by raising
its expectation of food and then delaying to give.
Its picturesque appearance when thus excited,
setting up all the long feathers of its crest, half
spreading its wings and striking keenly with its foot.
Its eyes are very beautiful. I long ago saw some of
these interesting birds, not only alive, but wild,
at the Cape, and was very glad to renew my acquaintance with it here. It appears very healthy
and full of spirit. The way in which, when excited,
it sets up the black quill-like feathers on its head,
which might well suggest the idea of pens, explain
the name of Secretary. *Serpentarius*, however, is a
better name.

The Cariama, from the campos of the interior
of Brazil. Several specimens, very lively and
almost comically tame, one in particular, which
pulled pertinaciously at the fringes of Fanny's dress.
On some signal or incitement from the keeper, they
all set up the most outrageous screams I ever heard,
throwing their heads back and opening their mouths
wide, as if on purpose to scream their utmost.

The Cariama *(Seriema,* I believe, is a more
correct form of the name). According to the
Zoological Society's list, this is an intermediate
form between the Bustards and the Cranes; in
general appearance it is much like a Bustard, with,
I think, a longer beak and with a crest of delicate
feathers on the head. It is a handsome bird.

There is a single specimen here of a second
species—Burmeisters Cariama—a great rarity.

The Chionis is placed in the Zoological Society

list, between the Plovers, Oyster-catcher, and 1878. Turnstone, on the one side, and the Curlews and Sandpipers on the other. According to Wallace ("Distribution of Animals") the *Oyster-catcher* is the nearest ally of the Chionis. The Sheathbill, Chionis alba, a very rare and peculiar bird, entirely white; in size, general air and appearance and actions, rather like a Plover, but with a very peculiar beak and face. It is a native of Fuegia and the Falklands: very rare in collections; I believe this specimen is almost if not quite the first which has been brought alive to England.

The pretty little burrowing Owls of the Pampas, three or four specimens, quite lively in the day-time.

June 25th.

Dear Minnie took me in her brougham to the Royal Academy; we spent more than an hour in the Exhibition, and saw it comfortably. Though it is not a first-rate Exhibition, there are several interesting pictures. I noticed and approved particularly the following:—

Figure or subject pictures (not portraits);—" The Gods and their makers" *(Long)*; girls of ancient Egypt working at little images of the animal gods— very clever and amusing; the expressions of several of the girls capital, especially one who is holding up—with a very comical satirical look—a little idol, and another who is holding a live cat for a model. I do not think it is quite equal to his Babylonian Marriage Market, two years ago, but it is very good.

1878. "A Worker in Brass," Antwerp *(R. Browning)*, very
well painted indeed. "After an Entomological
Sale—Beati Possidentes" *(Armitage)*, very clever;
the countenances of the lucky man who has bought
a great rarity and is immensely delighted, and of
the disappointed competitors, are capital. "Morning" and "Evening"—delightful groups of dogs, by
Ansdell.

June 26th.

Our luncheon party:—Lady Winchelsea, Lady
Eveline Finch Hatton, Lady Gwydyr, Miss
Burrell, two of the Miss Egertons, Mr. and Mrs.
Bontein, Mrs. Laurie and Miss Bailey, Sir John
Shelley, besides Isabel and ourselves. A very
pleasant party and good music. Lady Winchelsea
is very interesting.

June 27th.

Weather splendid, as it has been all this week.

We (Isabel with us), went to the afternoon party
given by Mr. Roundell and Mr. Rogers at the
Dulwich Gallery—most enjoyable: pictures, garden,
and many acquaintances.

June 28th.

We went to Lady Codrington's afternoon party,
given in rather a peculiar but agreeable way, in the
interior garden of Eaton Square:—much more
agreeable in such weather than an in-doors social
party.

Revisited with Fanny the Royal Academy.

"The Road to Ruin" *(Frith)*: a series of five small pictures, illustrating the career of a foolish young man through all the stages of dissipation and reckless extravagance, ending in suicide: very cleverly painted, and very painfully true to modern fashionable life. The idea of course is taken from the Rake's Progress, but the treatment is free from coarseness.

"Sympathy" *(Riviere)*—a delightful white terrier trying to console a child in disgrace—capital.

Landscapes:—"A Showery Day" *(Vicat Cole)*— delightful. "Wandering Shadows" *(P. Graham)*— most excellent; a piece of moorland mountain scenery in the Highlands with the lights and shadows on the rocky and heathy and mossy steeps. "The Alps at Rosenlaui" *(V. Cole)*—a subject very different from his usual ones, but very well treated. "In the Bernese Alps; a storm coming up" *(Sir Robert Collier)*—very good indeed. "The Timber Waggon" *(C. E. Johnson)*—a woodland scene, with a fine distance.

Of the portraits, I was chiefly pleased with some charming pictures of children, by *Eddis* and *Sant*.

———

June 29th.

Our dinner party:— Sir Joseph and Lady Hooker, Lady Mary Egerton, Lady Lilford, Lady Rayleigh (the dowager), Mr. and Mrs. Cyril Graham, Mr. Bentham, Mr. and Mrs. Clements Markham, the MacMurdos, Isabel, Edward, Harry, Arthur.

1878.   June 30th.

A great thunderstorm about the middle of the
day, and rain all the afternoon. It is rather
singular, this recurrence of a thunderstorm at the
same time as last Sunday, the whole of the inter-
vening days having been remarkably fine, bright
and warm. Indeed it was splendid summer
weather during the whole of that time, and most of
the days were hot, some uncommonly so; on Wednesday the 26th, the thermometer in London stood at
91 degrees in the shade. A fine week for the hay-
harvest.

_____

July 1st.

Our dinner party; our dear Bishop of Bath and
Wells, dear Sarah and Albert Seymour, Lord and
Lady Tollemache, Sir Robert and Lady Cunliffe,
Sir Charles and Lady Ellice, Mr. and Mrs. Wil-
loughby Burrell, Colonel and Mrs. Rowley, Isabel.

A very pleasant party. I was placed between the
darling Sarah and Lady Tollemache, who appears
to me a decidedly clever woman, and of some
learning.

An interesting conversation after dinner between
the Bishop and Lord Tollemache, on the subject of
Stonehenge and other British or Roman-British
antiquities. The Bishop had been much interested
in Stonehenge and Old Sarum, which he had lately
seen, apparently for the first time.

_____

July 2nd.

We went to the MacMurdos at Rose Bank,

where there was a numerous party assembled to
hear Mr. Brandram recite a play of Shakespeare.
" The Merchant of Venice " was the play chosen,
and it was extremely well delivered. The Shylock
in particular was admirable ; only now and then I
thought he was a little too loud—that it might have
been better if his malignity had been a little more
subdued in its expression.

July 4th.

Our dancing party, or rather my nephew Harry's
—for the notion was first started by him, and the
arrangements have all been carried out by him and
Isabel Hervey, under Fanny's sanction. It has
turned out very fortunately, and seems to have been
much approved by the parties principally concerned.
A charming bevy of very pretty girls.

July 5th.

The Aberdares with their son Harry and two
daughters, also Cissy and Emmie, and Harry
Bunbury dined with us. A pleasant little party.

Very agreeable conversation with Aberdare and
Norah. Aberdare has a very highly cultivated
and remarkably refined mind, with a great amount
of knowledge, and very sensitive feelings. My own
impression is, that he is better fitted for a life of
literature, or at any rate, of theoretical politics, than
for a career of active political struggles.

He talked delightfully about Richardson, Fielding, Smollett, Sterne, Miss Burney and other old

1878. authors. Advised me to read "Sir Charles Grandison," but with the omission of the story of Clementina ; said that the *secondary* characters in that novel, such as Charlotte Grandison are thoroughly natural, and not even antiquated in manners.

————

July 6th.

Fanny went with Harry to Sandhurst to see Susie and Cecil. Minnie with me to Cornwall Gardens, and saw Susan Horner, Katharine and Rosamond Lyell, Annie Pertz and Sophy Lyell, all newly arrived from Italy. Much pleasant talk with them.

Afterwards, we went to the South Kensington Museum and spent some time very pleasantly among the pictures lent by Lord Spencer. Much the most interesting of them are the historical portraits, especially of the Spencer family, many of which I knew by engravings.

————

July 7th.

A visit from my old tutor Mr. Matthews, who is now upwards of 80 years old : a quaint-looking, eccentric, but good little old man, who still nourishes very warm feelings towards his former pupils.

————

July 8th.

Fanny had a very pleasant letter from Clement,

dated " Invercargill, New Zealand, May 16." He 1878.
seems to be much pleased with what he has seen,
and to have travelled through nearly the whole
length of the two islands, and seen all that is
interesting. He describes the western coast of the
Southern Island as very wild and magnificent in
scenery, but very alarming and dangerous for
navigators. He had a good view of Mount Cook,
which he calls the Mont Blanc of New Zealand, it
is 12,200 feet high. He lost his way on a mountain,
and seems to have had rather a narrow escape, the
more so as the snow was falling fast.

Invercargill, from whence Clement dates his
letter, is between the 46th and 47th degrees of S.
lat. The climate appears to be rather severe, for
he speaks of heavy falls of snow while he was there.
It is true that the season was the winter of those
latitudes : yet most of the New Zealand plants
require shelter in an English winter, whence one
would infer a milder climate. Invercargill is not
far from Dusky Bay, where Tree-Ferns, and I think
a species of Areca, were observed by Menzies.

Our dinner party :—Lord Charles Hervey, Lord
Talbot de Malahide and Miss Talbot, Leopold and
Lady Mary Powys, Mr. and Mrs. Maskelyne,
William Hoare and Mrs. Hoare, Mr. White (of the
Savoy Chapel), Mr. Kinglake, Minnie and Isabel.
A very agreeable party. I sat between Lady Mary
Powys and Mrs. Maskelyne, the latter thoroughly
charming, the former lively and cheerful, abounding
in good humour and good nature.

The Holland House party—to which we went, a
party of six, in two carriages—Isabel Hervey,
Minnie and the Edward Goodlakes, besides ourselves. The day was perfect, bright, enlivening—
neither too hot nor too cold — and the scene
extremely pretty and agreeable ; those beautiful and
classic grounds animated by a multitude (not a
crowd), of gay and brightly dressed company.

_____

July 16th.

Another beautiful day.
We took leave of dear Isabel Hervey, who has
been a very pleasant inmate of our house ever since
the 21st of May, and went down to Barton : Susan
Horner and Annie Pertz with us. We had a good
journey, arrived safe and well, and found all well at
home. Thank God.

Isabel Hervey is a very pleasant, very engaging
and very loveable girl, very lively, bright and
cheerful, gay and fond of amusement, very sweet-
tempered with a remarkably warm and affectionate
heart. Her special pursuit is music, to which she
is passionately devoted. I take it for granted that
she is very skilful therein.

_____

July 18th.

My beloved wife's 64th birthday. Thank God,
she is in good health.

The Treaty of Berlin has been signed, the
Congress has closed, and the Plenipotentiaries have

returned to England. Thus we have the comfort 1878.
of feeling ourselves secure of peace, at least for
some time to come, and released from those
apprehensions and anxieties by which we have been
agitated for the last two years. How many years
the peace may last, it is a difficult question to
answer: for I cannot suppose that either Russia or
Turkey is pleased with the conclusion arrived at:
but it is at any rate unlikely that either will be
ready to renew the struggle till after some years.
We have certainly had a narrow escape; several
times since the treaty of San Stefano became
known, it has seemed almost impossible that peace
should be preserved: and I cannot deny that, as
far as I can see, honour is due to Lord Beaconsfield and Lord Salisbury for their management of
the negotiations, by which they have preserved
peace while upholding the dignity and importance
of Britain.

Perhaps, after all, Bismarck has contributed more
than anybody else to preserve Europe from a
general war during this dangerous time.

The Convention between England and Turkey,
by which we have obtained possession of Cyprus,
undertaking in return the defence of Asiatic Turkey
in case of aggression, and accepting a sort of
responsibility for the good government of that
country—this is indeed a bold, a daring, a startling
measure. On its first announcement it seemed
almost terrific. It is an extreme instance of Lord
Beaconsfield's love of grand and startling *effects*—of
grands coups de theatre. But I hope it may in the

1878. long run prove less dangerous than it appeared at first : and it is no doubt a large and statesmanlike scheme. If we govern Cyprus as we ought to do, there is some hope that it may in time become a pattern to other eastern governments, and have a wholesome influence upon them.

_____

July 20th.

Yesterday I read Lord Beaconsfield's speech in the House of Lords, explaining and vindicating the Treaty of Berlin;—a very able and powerful speech. He makes out a very good case for himself and his colleagues.

_____

July 24th

A violent thunder-storm in the afternoon, with very heavy rain, which (the rain) continued with intervals till night. It was very welcome. Hitherto, ever since we came down, the weather had been brilliantly fine—hot, and dry without a drop of rain; and so, by the accounts I have had, it seems to have been here and hereabouts, for nearly a month past. The storms which visited London and its neighbourhood on the 23rd and 30th of June, and in the first week of July, did not reach this part of Suffolk. The weather has been splendid for hay-making, and *that* crop, in my park, has been the finest (Scott says) *and got in* in the shortest time of any that he remembers at Barton. The Wheat also looks very promising; but some other crops, as

well as the pastures, were beginning to suffer from 1878.
the continuance of heat and drought. The gardens
also will be much the better for the rain.

July 31st.

A great change in the weather; the last four or
five days very cold, though not rainy except
yesterday. Yesterday the Assizes at Bury; I, as
usual, Foreman of the Grand Jury. Not much
business; sixteen prisoners for trial, few of them for
very serious offences, and only one civil cause. I
dined with the Judges—Justice Thesiger and Baron
Cleasby. Thesiger (a son of Lord Chelmsford)
is one of the Justices of Appeal, and is hardly above
40, being the youngest of all the Judges. He is
agreeable, and evidently an able man; with an
expressive and rather fine countenance, which
appears to indicate firmness and power. Baron
Cleasby is an oldish man, lame (perhaps by
accident), with a mild, benevolent countenance.

August 3rd.

The Wheat harvest is now begun pretty generally
in this parish, except on one or two farms where the
soil is heavier or colder than in the rest; and it is
hoped that the crop will be tolerably good.

August 6th,

Scott brought me from the Home Farm, a bunch
of ears of Wheat, which he said were about the

1878. finest he had seen grown here. They are of a *Red* Wheat, and of a sub-variety, with whitish chaff (*glumes*).

Glorious weather; the harvest going on superbly. Scott tells me that in Bury market, on Wednesday, the accounts were excellent, and the new wheat which was sent in was in the finest condition. Beautiful as the weather is, the garden, lawn, and park, give no sign as yet of suffering from drought, The foliage of the trees is very fine, and the Catalpa and other ornamental flowering kinds are in great beauty.

Our house full of company ever since we came from London; a family party mainly,—many young people, much gaiety and merriment. Very pleasant, but not favourable either to business or study.

Personages :—

Susan Horner; Annie Pertz; Cecilia Bunbury; Emily Bunbury; Minnie Napier; Sarah Seymour; Charlie Seymour; Montague and Susan MacMurdo; Emily MacMurdo; Lady Head; Miss Wilmots; Mr. Courtenay and Lady Muriel Boyle; Mr. and Mrs. Livingstone; Willoughby and Mrs. Burrell; the Barnardistones, and Lady Louisa Legge,, and the Godwin-Austins; Susan Horner, looking well, though very thin; in very good spirits and full of animation; quite the same as ever, in her prodigious fluency of language and great power of expression, in her eager

activity of mind, her warmth of feeling, strength of 1878.
friendship, and energy of intellect. She does not
seem to me to have grown at all older in the four
years since she was last in England.

I like the Burrells very much; the lady is
particularly interesting, and very agreeable. Lady
Louisa Legge very clever and amusing.

LETTER.

Barton Hall, Bury St. Edmund's,
August. 15th, '78.

My Dear Katharine,

      I have committed to Susan's care several
packets of Charles Lyell's letters, which are very
much at your service for your work, and in which I
dare say you will find a good deal that is characteristic and interesting. They are roughly
arranged chronologically, but there are a few
undated, for which I dare say you will find dates
from internal evidence. I was very sorry to say
good-bye to dear Susan, her visit here was exceedingly pleasant to me, and she looked in good
spirits and happy, and was (I hardly need say)
extremely agreeable. 1 hope she will not tire herself by the journey to Scotland. Annie is a very
nice girl indeed, and appeared very happy with the
other young ladies; indeed they all looked a very
bright and joyous party, and were very pleasant to
look at. Annie worked very hard at her painting
while she was here, and has made a most capital
copy of "The two Miss Hornecks." You will have

1878. heard from Fanny and Susan all about our party in the house for the last month; it has been very pleasant independently of the family elements. It has been a very pleasant time, but I have been very idle. At first the enjoyment of idling and strolling about in the delicious weather, the delight in the flowers and trees, made me a mere lounger; and latterly, the racket of company, dissipated my thoughts, so that actually I read better in London than I have done since we came home. I get on slowly with Green's "History of the English People," (the large edition);—not that it is not an interesting and valuable book, but it is stiff reading and requires much attention. I am reading with a good deal of pleasure Leslie Stephen's "Hours in a Library"—a series of critical sketches of authors and books, mostly of the last century; some of those I have already read are very good. The trees and flowers and grass here are still in great beauty, though the Catalpa and others are passing away. I feel very unwilling to leave home, though I shall be very glad to see dear Susan again in London, and not sorry to get some sea air on the South coast, as well as, possibly, one or two new ideas. I hope you are passing your time pleasantly in Scotland; do you botanize at all? The Austens admired our trees in the arboretum, and, indeed, they are in beautiful foliage this year, and in the hot days it was very pleasant lounging under their shade.

The two or three thunder storms we have had lately (with heavy showers) have done no mischief. The harvest is going on briskly, and I believe much of the

wheat is carted, but it seems to be thought that the 1878.
crop will not be so good as was at one time hoped.

(August 16th). Our party is now reduced to Lady
Muriel Boyle, Edward, Clement, Arthur, and Mimi
MacMurdo. Lady Muriel leaves us to-morrow.
Pray give my best remembrances to the Miss
Lyells, and my love to Rosamond, and believe me
ever,

Your affectionate brother,

Charles J. F. Bunbury.

JOURNAL.

Clement arrived yesterday evening, very lately
landed on his return from New Zealand and
Australia. He is looking very well; and seems
in good spirits, has plenty to say about the strange
countries he has seen, and tells it fluently and well.

He has observed well, and gives us very interesting particulars about the forests and the gigantic
trees of Victoria and Tasmania. He actually saw
and measured, in Victoria, trees of Eucalyptus
upwards of 300 feet in height; not standing trees
of which the height was guessed or calculated, but
felled trees which were actually measured when
lying on the ground. These were the largest he
actually saw, but he was told, and believed, that
others of 400 feet had been measured. Those
Victorian trees however had trunks very slender in
proportion. They grow chiefly on the outer and

1878. lower ranges of the mountains in Victoria. The
great trees of Tasmania grow especially in the
Huon district, not very far distant from Hobart
Town. They appeared to him to be of greater
girth, though not of greater height, than those in
Victoria; but in many of them the apparent bulk
seemed to be increased by buttress-like projections.
In the New Zealand forests he was struck, not
so much with the great height of the trees, as with
the luxuriance and dense entanglement of the
vegetation and the gigantic woody climbers, which
must be like those of tropical countries. He made
acquaintance with Dr. Hector and Mr. Mantell, and
received much help and much interesting information from them. He remarked, both in Australia
and New Zealand, the very rich and well-arranged
and valuable museums and public libraries in many
of the towns, even those of secondary importance.

Clement's return voyage was from Adelaide in
Australia, by Aden, Suez, and the Mediterranean;
they saw no land between Adelaide and the Island
of Socotra at the entrance of the Red Sea. Their
voyage so far had been extraordinarily rapid, and
the weather most favourable; but near Cape
Guardafui they were in great danger of shipwreck,
for the weather was so thick for some days that they
could get no observation, and a strong current
carried them quite out of their reckoning. It is
a coast region notorious for shipwrecks. The ship
in which Clement was, made the voyage from
Adelaide to Aden in 21 days.

Clement tells me that pheasants, both the common European and the Japanese Phas. versicolor, 1878. have been successfully naturalized in New Zealand, and are now common in the woods there. He was struck (as I should have expected from what I have read), with the scarcity of native birds.

***

August 24th.

Up to London, with Mimi (Emily) and Arthur MacMurdo ; the last who have remained of our party of guests.

We found Susan Horner and Annie Pertz established at 48, Eaton Place.

***

August 26th.

Fanny, Arthur, and I went to the Haymarket, to see "Our American Cousin,"—Mr. Sothern, in Lord Dundreary—and I hardly remember to have laughed so much in my life ;—certainly not since the days of my youth. Dundreary is quite irresistible. It is really a satisfaction to find that in my 70th year I can laugh so much. Arthur's contagious merriment and delight added to the enjoyment.

***

August 27th.

Had the pleasure of a visit, though a flying one, of only a half-an-hour, from dear Rose Kingsley. She seems well and as warm-hearted and pleasant as ever. She told us also a pleasant piece of news—that our dear Isabel Hervey is engaged to be married, with the approbation of her parents, to

1878. Mr. Calliphronas, a semi-Greek gentleman, whose
mother was Miss Charlotte King, sister of Locke
King and of the present Lord Lovelace. I heartily
wish that all may turn out happily for Isabel, for I
am very fond of her.

(*August* 31*st.*)—Fanny has had a charming letter
from dear Isabel, delightful from its tone of over-
flowing happiness, and full of the beautiful candour,
simplicity and kindness of her nature.

LETTER.

48, Eaton Place,
August 27th, '78.

My Dear Katharine,

I have for some days been meditating to
write and tell you of the new acquisitions, brought
by Clement, to our Fern house at Barton. He has
brought home several (I think seven or eight),
plants—that is living trunks or logs—of Tree
Ferns, which are now under Allan's care; they are
quite alive, and I hope may flourish, though we
shall hardly have room for all of them. They are
of two species—Cyathea dealbata, which is clearly
indicated by some withered remains of fronds,—and
Cyathea medullaris I presume, but this latter has
no remains of leaves. There were some smaller
kinds which had been packed with them, but these
had died of heat on the voyage. Moreover these
ferns were packed in moss, and as of course this
was New Zealand moss, I thought it worth while to

have it unpacked carefully, and have roughly sorted and dried a good lot of it, so as to have it ready for future examination. I could see at once that most of it was different from European mosses.

Thank you for your interesting remarks on Charles Lyell's letters. Almost everything he wrote was of permanent value. I am very glad you have got copies of his letters to Dr. Mantell.

Susan and Annie are not here just now: they left us before we had been here 24 hours, to go to the Darwins, but we expect them back this evening. Susan's energy is wonderful, as indeed it always has been, and though she is very thin, she does not seem to me to be looking ill.

Annie's copy of the Miss Hornecks is most beautiful—really almost perfect.

Your stories of Charlie Lyell are very interesting, indeed he is an interesting little fellow.

I have not time to write more just now, but pray give my love to dear Rosamond and to Leonard and Mary when you see them, and my kind regards to the Miss Lyells, and believe me ever,

Your affectionate brother,

CHARLES J. F. BUNBURY.

JOURNAL.

August 29th.

We went out with Susan and Annie, and saw the Bridgewater Gallery, which, strange to say, I had never seen before. It is a beautiful and very in-

1878. teresting collection, and contains a great many really
first-rate pictures; in particular, some most delightful *Raphaels*; *Titian's* famous " Four Ages of Man "
(charming) ; *Titian's* " Venus Anadyomene," and
several others of his best known mythological
pictures (most of them were formerly in the Orleans
Gallery, and I knew them well from engravings) ; a
Cuyp, to my thinking, the finest I have ever seen.

———

August 30th.

We, with Susan, Annie, and Arthur, visited the
National Gallery. It is now a really noble collection of pictures, exceedingly rich, well arranged, and
placed in fine and spacious rooms, where it can be
seen to great advantage. Susan Horner, who is
familiar with the great Italian Galleries, was very
favourably impressed by this.

There are some new pictures, this year ; a
striking and fine one by *Ward*, a view of Gordale
Scar in the Craven district of Yorkshire ; a grand
scene of towering and frowning precipices, with
cattle and deer assembled on the level space at their
feet.

———

August 31st.

There was a great fire on the night of the 28th, in
a timber yard, in Buckingham Palace Road, and it
is not out yet; when Arthur went to look at it this
morning, he found it still burning considerably, and
several engines at work upon it. The firemen told
him they did not expect to be able to extinguish it

completely yet for three or four days; but it is 1878.
enough subdued to obviate the danger of its
spreading.

The weather, during these days that we have
spent in London, has been excessively stormy,—
frequent and sometimes violent thunderstorms,
frequent and heavy showers of rain. Deplorable
for the harvest, and accordingly I receive melancholy accounts from Suffolk.

Dear Susan Horner and Annie Pertz set off on
their return to Italy. We three—Fanny, Arthur
and I—went down to Portsmouth by the 2.35 train,
and then drove to the Albert Seymours, St. Elmo,
Clarendon Road, Southsea. A very fine day. A
beautiful country, from Guildford by Godalming,
Liphock, Petersfield and Havant; variety of surface
and covering, rich woods, heath, corn lands, hop
grounds, chalk downs. From the narrow chalk
ridge of the North Downs, which we cross at
Guildford, we go on over the Upper Green Sands,
Gault, and a great extent of the Lower Green
Sand, till we come again to the Chalk near
Petersfield, where the North and South Downs
unite: and a little short of Havant find ourselves on
the Tertiaries.

We remained at Southsea till yesterday, doing
very little, but enjoying the delightful society of
dear Sarah Seymour and her husband and their
darling little boy Charlie. This was a great
pleasure. We remained in their house till the 7th,
then removed into an excellent hotel—the Beach
Mansions Hotel—situated on the very beach, close
to the sea, at the extreme E. end of Southsea, near
to Eastney Barracks. Here we were very comfortable, had full enjoyment of the sea air, and saw
the Albert Seymours every day. The weather was
most beautiful almost the whole time we remained
at Southsea; the sea like a mill-pond, and the sky
generally cloudless till after the middle of the day;
the afternoons sometimes cloudy, and the evenings
generally chill though fine.

A good deal of animation in the sea-views from
our windows; plenty of ships and boats of all sorts
and sizes, once or twice a turret ship passed in sight
but not near enough for me to see clearly its
structure; it appeared a strange, uncouth monster.
Several times we were entertained by seeing the
forts on the shore, or gunboats, firing heavy guns at
a floating mark.

One day we took a drive to Alverstoke and round
by Haslar Hospital, crossing and re-crossing the
great Harbour by the Floating-bridge. An animated
and interesting scene in the Harbour;— shipping
innumerable—the great ships—the *Duke of Wellington*
(the flag ship), the *Victory*, the *Vincent*—enormous
troop-ships—steamers and boats of all sorts, swiftly

passing in all directions. Another day we went to 1878. Eastney Barracks and Forts, situated on the shore to the east of Southsea Common. The Barracks very large. A considerable piece of ground adjoining laid out in allotments which are cultivated by the soldiers.

We twice visited Porchester Castle, a ruin of considerable dignity for size and picturesque appearance. It stands on a projecting, but not high tongue of land, almost surrounded by a branch of Portsmouth Harbour; and when the tide is in, the castle is seen to great advantage across this sheet of water. The conspicuous part of the castle is the square Norman keep, which rises high above the rest. "The outer court, or ballium, is formed as at Pevensey, by the walls of the original Roman fortress."—(Murray's Guide). This court occupies about nine acres; the walls are from eight to 12ft., thick, and about 18ft. high; they are built chiefly of flint, with little or nothing of the usual Roman tiles. "The bonding courses of the walls are formed of a coarse limestone, which is also used at Silchester, and the composition of the cement is the same at both places."—(Murray). The towers in this outer line of wall are round, and remind one of Burgh Castle, but they are hollow, not solid as *there*.

The botany and geology of Southsea, absolutely *nil*, except as to the flowers in the gardens, which indicate the mildness of the climate. Passiflora cærulea covering the fronts of the houses with a most luxuriant growth, flowering abundantly, and often bearing its beautiful fruit of a bright, rich apricot colour. Fuchsias of many varieties, very

1878. large and flourishing in the open ground; particularly
fine and luxuriant in a garden at Porchester, in
which also a Yucca, very fine.—Hibiscus Syriacus,
very large and flowering profusely in many of the
gardens.

Mrs. and Miss Ellice were staying at Southsea,
and we saw them frequently, very pleasant.

_____

September 18th.

We left Southsea (of which I was considerably
tired) on the 14th, went by railway to Romsey, found
Mr. Smith's carriage waiting for us, and drove three
miles to his house at Embley, where we found a very
kind welcome.

Romsey appears a neat and cheerful town. Immediately on going out of it to the West towards
Embly, we cross the river Test, a beautiful clear
stream, famous for Trout; and we pass by the
entrance of Broadlands, famous in Lord Palmerston's
time. Our kind old friends Mr. and Mrs. Smith,
received us with great cordiality, and though _he_ is 84
years old, and she 80, and they are naturally feeble
from age, their mental powers seem but little im-
paired.

Embley is a beautiful place. The house a large
and handsome one, of red brick, in the style of James
the First's time, stands high; the park slopes down
from it into a green valley, and rises again into
beautifully wooded hills, which seem to fold in
graceful curves round the lower grounds. From a
walk which runs along the brow of the hill opposite
to the house there is a very fine view looking west-

ward over the park, and the New Forest, to a range 1878.
of considerable hills which bounds the prospect on
the W —The slopes of the open ground in the park,
the rich woods, and the disposition of the scattered
trees, make a delightful scene.

Trees, of most kinds, grow very finely here,
especially oaks, sweet chestnuts, and Scotch pines,
of all which there are superb specimens in abundance in the park and woods. But the birch is, as
Mr. Smith expressed it, the *weed* of this country.
Many exotic trees and shrubs grow luxuriantly; the
Ericaceæ most especially. The Rhododendrons of
Embley I had often heard of, and I now saw enough
to satisfy me that they deserve their fame. They
are of course not in flower at this season, but their
height, density and luxuriance of growth, and the
extent of ground they cover is surprising. One
drives actually for miles through a continuous underwood of Rhododendrons, as high as any common
Laurels, and so dense that it would be impenetrable
without cutting. Kalmias, and all that family of
plants, seem to thrive as well as the Rhododendrons; and many exotic Conifers and other
trees also.

I noticed particularly; deciduous Cypress (near
the house),—much the tallest and largest tree of the
kind that I have ever seen;—with a robust bare
trunk, rising to (I should suppose) at least 15ft.
before a branch, and 8½ft. (I am told) in circumference;—a fine, dense pyramidal head of beautiful
foliage. There are some other uncommonly fine
trees of this kind in the grounds, but this is the most

1878. conspicuous. Pinus insignis—one of the largest and finest I have ever seen.

Cedars, very many, and several very noble.

Tulip tree—the largest I have ever seen.

Magnolia tripetala—very fine and healthy.

Bamboo—one of the Chinese species I presume—thriving in the open ground; forming a dense clump but upright and stiff in growth, not gracefully arching like those in Brazil.

Soil sandy—Bagshot sand I presume—mixed towards the surface with much of black heath-mould.

Heaths—all our three common species—in great profusion and remarkable luxuriance in the wider and more open parts of the woods; the flowers now nearly past but not quite.

Narthecium; Mr. Smith and Mrs. Coltman told me it grows plentifully in one part of the park or wood.

Mosses and lichens evidently very abundant and of vigorous growth. I gathered Neckera pumila on trees near the house, and I dare say I might have found some other good mosses if I could have had a solitary walk. The ground under the trees in the pleasure ground carpeted with Mnium hornum, which I observed last year equally abundant in the same situation at Alfreton.

————

September 24th.

Barton.

We went from Embley to London on the 17th—remained at Eaton Place the 18th and 19th—during which time we were heartily glad to see dear Minnie, newly returned from a rapid tour with her brother in

the Pyrenees, and looking the better for it. The 1878.
18th, Arthur returned to his tutor's at St. Leonards.
We had a visit from Mrs. Horton, with Freda and
her betrothed, Sir Lambton Loraine—all looking
very radiant.

The 19th, we went down to the Millses at Stutton
—by railway to Manningtree (about two hours),
thence a long drive to Stutton. The 20th, we
remained quiet there, with our dear old friends, who
are as cordial and kind as ever. Mr. Mills is (I
believe) 86 years old, but looks very well for that age,
and seems to have his mental faculties in good
preservation. Stutton is quite a gem.

The 21st, we returned home and found all well
here. Thank God.

While we were at Southsea there occurred the
most dreadful catastrophe, the running down and
sinking of the "Princess Alice" steamer by another
steamer in the Thames, with the loss of more than
600 lives. I have seldom heard of so sudden and
awful a catastrophe: indeed, except in the case of an
earthquake, it rarely happens that so many perish at
once. And what perhaps makes it more shocking is,
that the victims were returning from a day's gaiety
and amusement,—they were fresh from the excitement of a day of pleasure, when they were suddenly
swept into eternity. Almost immediately afterwards
came another terrible disaster—the great explosion in
a colliery in South Wales, with a frightful loss of life.
Colliery explosions unfortunately are very common
events, but I hardly remember to have heard of one
so destructive as this.

1878.   October 4th.

A Riddle.

Why was not London illuminated for the Triumphal return of Disraeli?—Because they thought the *Jew* (*le jeu*) was not worth the candle.

October 5th.

Clement here since September. His remarks on what he saw in Australia and New Zealand. The beauty of the Tree-ferns; especially the marvellous effect of their delicate fronds when seen against the light—of sunset for instance—in a ravine, or an opening in a forest. The vegetation in different stages of height as he saw it in Victoria;—Adiantums and other low growing delicate Ferns covering the ground, and reaching to his knees, or less,—overhead, the Tree-ferns forming a canopy;—and far over these, again, the giant trees, 200 or 300 feet high.—Black Swans—he saw a very large flock of them in Tasmania;—when on the wing, the white colour of part of their wing-feathers very conspicuous.

White Cockatoo, very numerous in New South Wales, and very destructive to the crops, he has seen trees growing near corn fields, absolutely covered with them.—In Victoria, the gold bearing veins of quartz have been worked with success, by vertical shafts, even down to 1,200 feet below the surface, and the miners believe that they will be found productive as far down as they can be reached.

Clement has brought a great many photographs, very interesting illustrations of the scenery of the countries he has lately visited. Many of the forest 1878. views admirable—most interesting; the best representations of Tree-ferns I have ever seen; also of rocks and of the wonderful terraces of siliceous deposit at Lake Rotomahana. The photographs of Tree-ferns and of underwood, forcibly remind me of Brazil.

He has noted the size of a famous tree not very far from Melbourne, Victoria; it is 425 feet in height, 62 feet in girth.

October 11th.

We seem likely to be entangled in another Afghaun war;—sure to be unsatisfactory, and likely enough to be disastrous. There have, within these few days, been three important letters in the *Times*, from Lord Lawrence, Lord Grey, and Sir Charles Trevelyan,— arguing against such a war, contending that there is no sufficient provocation or real necessity for it, and that it would do us more harm than good. Montagu MacMurdo thinks otherwise; he holds that to take vigorous measures against Shere Ali, to punish him for the insult, is absolutely necessary to preserve the respect of the natives of India. He does not think, however, that there will be war; he believes that Shere Ali is so much detested by his own people, that they will turn against him, and kill or expel him, rather than follow him in war against us. I wish it may be so. MacMurdo fully admits the extreme difficulty of the Afghaun country, and the necessity of making most ample provisions and

1878. preparation for the campaign. MacMurdo thinks
that, even if the Afghauns resist in earnest, and we
have to subdue the country, it will not be necessary
to hold it permanently, but only to occupy it for a
time and then withdraw.   I am afraid I do not see
how this can safely be done.

————

October 12th.

Clement has brought from New Zealand, a very
fine specimen bird (stuffed) of the rare Night Parrot of
New Zealand (Stringops habroptilus), which he has
had mounted in England, and which is now here.
It is a much larger bird than I expected to see;
indeed he says it is an uncommonly large specimen,
and that Doctor Hector and others told him it was
the largest they had seen.   At first sight (at least
when *set up*) it has rather the look of a bird of prey,
than of a parrot.   The colouring too, is peculiar,—
mottled all over with black and yellow on an olive-
green ground.

————

October 13th.

Thermometer in night down to within three deg.
of freezing point; a splendid morning.   The weather
has been beautiful, almost without interruption, from
the very beginning of this month ; the sun often
powerful in the middle of the day and in places
sheltered from the wind; the nights fresh, and
latterly rather cold but no frost yet.   The autumnal
colouring of the trees beautiful.

————

A house full of very pleasant company ever since the first of this month.—Charlotte and Octavia Legge; Mr. and Mrs. William Hoare; Admiral Spencer; John Herbert; John Hervey; Mrs. Ellice and Helen; Augustus and Lady Mary and Miss Phipps; Mr. Dobree; the Goodlakes; the Mac Murdos (Montagu, Susan, and Mimi); William Napier; Katharine and Rosamond; Cissy and Emily; Sir Robert Cunliffe; Mrs. Frederick Campbell; and my nephews Clement, Harry, and George;—these of course, not all at once, but some coming as others went away. Now all are gone except Cissy, her daughter and her son George.

Most of these are old friends.

Sir Robert Cunliffe (with whom we became acquainted at Lord Hanmer's last autumn) is a remarkably agreeable, cultivated, accomplished man.

October 16th.

Dear Isabel Hervey was married yesterday to Mr. Cyril Calliphronas.

October 17th.

Read in the *Times* a very long and very important letter from Sir Bartle Frere, on our relations with the Afghauns. It is not actually new, having been written upwards of four years ago, before he went to Africa, and before the present "difficulty" began; but it is wonderfully appropriate to this crisis; and coming from a man of such eminent ability and

1878. wisdom as Sir B. Frere, and of such long experience
in those countries, it is certainly of great value and
importance. I do not feel, however, that in one
reading I have thoroughly mastered the sense of it,
and am not sure that I always rightly understand his
meaning. It certainly seems that he does not take
the same view as Lord Lawrence, or those who
would pass over the supposed insult without taking
any notice.

October 18th.

Finished reading the Memoir of Bishop Ewing;—
a well-written record of a very beautiful character.

October 27th.

We returned yesterday from a two days' visit to
Lord and Lady Walsingham, at Merton, in Norfolk;
—they are new acquaintances, yet we found the
visit remarkably agreeable. Lady Walsingham is
quite charming; very handsome, with peculiarly
fascinating manners, and plenty of intelligence; has
lived much abroad, and known a great variety of
people. Her manners, indeed, are perhaps a little
foreign, but they are singularly winning, and I have
every reason to speak with warm gratitude of her
attentive kindness to me. Indeed, if first impressions
do not greatly deceive us, she is really a thoroughly
amiable and kind-hearted woman. Lord Walsingham
also is very agreeable. He is a zealous naturalist,
devoted to Entomology in particular, and, I believe,
deeply learned in it, but very acute and observant in
other branches also of the science. He showed me

a small part of his collection, principally of minute 1878.
moths (Micro-Lepidoptera, to the study of which he
has especially devoted himself) most exquisitely
arranged and mounted. He says that the collection
of insects which he brought from North America,
will take him several years to arrange and describe.
He travelled for two years in the western parts of
that continent—California, Oregon, and British
Columbia,—collecting assiduously, and, as he is a
keen sportsman as well as a naturalist, his collections
are not confined to insects. His museum is crowded
with spoils of the chase; two stuffed bears, in
dancing attitudes, face you very comically as you
enter; and all round the walls are heads, horns,
tusks, and skins of Bisons, Deer of various kinds,
Bears, Boars, and large birds. He gave me a
quantity of dried plants which he had collected in
North-west America,—the greater number of which
I recognised as having been already given me by Rose
Kingsley, but several are new to me.

The weather on the 24th was excessively bad —
raining and blowing furiously all day. I spent a good
part of the morning with Lord Walsingham, in his
museum; and in the afternoon Lady Walsingham
showed us the house. The park is of great extent;
rather flat, (as is natural in Norfolk) but not without
undulations, and very pleasing from its fresh, open,
breezy, heath-like character, well variegated with
wood, with a fine growth of fern, and here and there
great trees. The woods near the house are beautiful;
natural and exotic growth mixed in a very agreeable
way, for under the great oaks and beeches, and

1878. amidst the luxuriant brake-fern, have been planted a
variety of foreign Conifers and other exotic trees and
shrubs, some of which are now of very fine growth.
The Rhododendrons may almost bear a comparison
with those at Embley. The soil is very light,
seemingly almost sandy, yet the oaks and beeches are
very large and fine, — as indeed they are about
Windsor. On the 25th (a fine bright day, though
cold), Lady Walsingham took me out in her pony
carriage, to show me the woods and the grounds
within reach of the house—a delightful drive.

Lady Walsingham, I understand, is daughter of
the Mr. Locke who was drowned in the Lake of
Como, and about whom I remember to have heard
much talk when I was a boy.

The company at Merton were:—Lady Alfred
Hervey and her very pretty daughter; the Edward
Goodlakes; Mr. and Mrs. Augustus Tollemache;
Lord Alexander Paget; Miss De Burgh; and Mr.
Cheney.

The clergyman of the parish came to dinner on
the 25th, he is Mr. Crabbe, grandson of the great
poet.

There are some fine well-grown Wellingtonias here,
some of the best I have seen.

Lilium giganteum (which I have not seen elsewhere) has been planted in several places in the
woods, and thrives wonderfully, growing eight or ten
feet high, and bearing (at this season) very large
capsules. It seems quite hardy. Its leaves are
peculiar, being so much broader than those of other
Lilies.

Weather very cold since we returned home. Every day since the 29th of October, the thermometer has descended below freezing point; on the 30th it was down to 25 deg., and there was a furious *snow*-storm —real snow,—in the morning.

Susan (Susie), William Napier's second daughter, and the wife of my nephew Cecil, died on the 6th, of congestion of the lungs. It is a terrible blow to her hasband, who was devotedly attached to her, and to her father and mother, who perfectly adored— idolized—her; and it is grievous to all of us; for though I never really knew her so thoroughly as might be supposed from the relationship, she was one who could not be known without being loved. She had a very lovely face, in a delicate, refined style of beauty, with which I believe her character and disposition were perfectly in harmony. No human being, I should think, could be more pure-minded, more meek and gentle, of a sweeter temper, or of warmer affection towards her own family.

Cissy has sent us a most interesting and touching account (which she received from Sarah Craig) of Susie's last hours, her last farewell to her parents and husband, sisters and servants—a death-bed of wonderful moral and religious beauty.

Before we received the sad news of Susie's last illness, we had had (as usual at this time of year) a very gay week; our house was filled with a merry

1878. and lively young party for the Bury Ball. There
were Lady Winchilsea, her daughter Lady Evelyn
Finch Hatton, and two sons; Mrs. Laurie and her
niece Miss Bayley; the Barnardistons, two of their
daughters and one son; Sir John and Miss Shelley;
Captain Crosbie, Harry Bruce, Charlie Napier,
Cissy and Emily, George, Willie, Arthur, and
latterly, Edward. A very pleasant party.

Lady Winchilsea (the Dowager—stepmother of the
present Earl) is a very interesting and attractive
woman; very handsome for her age (nearly 50),
noble and dignified in person, with most winning
manners, and a highly-cultivated mind. I sat next
to her at dinner every day, we talked on a great
variety of subjects, and I found her always
agreeable.

Her daughter is a very agreeable and interesting
young woman.

Sir John Shelley and his sister are very agreeable
young people: he has a cultivated mind, and his
countenance and manners are remarkably pleasing;
his sister is very handsome.

November 23rd.

Poor Cecil came to see us for 24 hours, arriving in the
afternoon of the day-before-yesterday, and going away
yesterday afternoon. It was a very sad but very interesting visit. He feels his bereavement with deep
sensibility, but bears it with a manly, christian and
noble spirit; no parade of stoicism, but a gentle and
firm resignation. He told us many interesting and
touching anecdotes of her last hours; it was indeed,

a marvellously beautiful death; she herself repeated 1878.
again that it was a "happy death, a delightful
death;" she was free from physical suffering, perfectly clear-headed, and in a most heavenly state of
mind. She took leave of all who were dear to her,
and addressed a few words to each of them, with
a clearness, a strength, a readiness, and force of
expression which (Cecil said) were perfectly astonishing to him; she seemed inspired.

LETTER.

Barton, Bury St. Edmund's,
November 25th, '78.

My dear Katharine,

I was intending to write to you nearly three
weeks ago, when the sad news of dear Susie's death
came in my way. You will have heard of course
from Fanny all the particulars. Poor dear Cecil
came here last Thursday, and stayed 24 hours;—a
very sad but very interesting visit. Poor fellow, it is
a sad lonely lot for him, but he will find a comfort in
reflecting on her beautiful and happy life, besides
the great consolation of remembering that he was a
truly devoted husband.

This loss is most grievous to him, but a terrible
one also to her father and mother, her brother and
her sisters. She seems to have had a wonderful
influence over all.

With the exception of this flying visit from Cecil,
we have been (for a wonder) quite alone since the
18th, but you may suppose we always find plenty to
do.

1878. So the Afghaun war has actually begun, and has already cost us the lives of two valuable officers. It is a sad business; I wish we may have sufficient justification for engaging in it, but I fear it is very doubtful. It is an awful responsibility that rests on the Government; and it is difficult for any one not thoroughly versed in Indian affairs to form a positive judgment, when one sees such differences of opinion among those who know these countries best.

The news of dear Susie's danger came to put a very sad ending to a very gay and lively party, who had kept us in a state of agreeable excitement for a week. They were very pleasant, both the younger and the elder members of the party.

The week before, we had spent three days very agreeably in a visit to Lord and Lady Walsingham at Merton in Norfolk.

Lord Walsingham very agreeable, a zealous naturalist, and especially entomologist, quite devoted in particular to the study of the minute moths, (Micro Lepidoptera) of which he showed us part of his collection, most beautifully mounted. He travelled for two years in California and Crogen, and brought home a collection of insects, which he said it would take him several years to arrange. He brought also a quantity of spoils of the chase—heads, horns, skins of Bears, Deer, Bison, &c., with which the walls of his museum are covered.

Give my love to dear Rosamond,

Ever your loving brother,
CHARLES J. F. BUNBURY.

JOURNAL.

We received the good news that dear Sarah Sey- 1878. mour has been safely delivered of a boy, and that mother and child are doing well. It makes us very happy.

———

Last week, again, a pleasant, cheerful social party; —the Louis Mallets, Lady Head and her daughter, Lady Rayleigh (the Dowager), Willoughby and Mrs. Burrell, Mr. Eddis;—for a part only of the time, Frank Lyell and Edward;—and for one day only, Mr. Lott and Mr. Livingstone. Several of our neighbours dined with us. Now we are again alone. Louis Mallet is (as I think I have already noted in this Journal) a man for whom I have a high respect and esteem; I consider him as a wise man and a very good man. Not that I think him always right in his political opinions; he is too Radical for me; but he is a thoroughly honest, earnest, true-hearted man, generally candid and moderate, always truth-loving and open to conviction. Of course everybody is at present interested in the war in which we are engaged. There was a good deal of difference of opinion among our guests. Louis Mallet objects strongly, on principle, to it;—thinks that it is un-necessary, and therefore unjust; that it might, by a judicious and steady policy, have been avoided without any danger to our Indian dominion; but he

1878. admits that, in the state to which things have now
been brought, there is nothing to be done but to
fight it out, till we have established a good foundation for peace.   I quite agree with him on this last
point; and I am now become a convert to his first
position as to which I was lately rather doubtful.
What I admire especially in Louis Mallet is, that in
his political judgments he seems to be always
guided by high moral considerations, not by those
of mere expediency.

In the meantime, our troops have been gaining
brilliant successes in the Passes, and it seems to be
generally admitted that the military operations have
hitherto been carried on with skill, ability and
courage.   I am very glad of this, not only for the sake
of our brave soldiers, but because rapid successes
seem to promise that the war itself may be short-
lived.   Louis Mallet thinks it very probable that we
may soon succeed in overthrowing and dethroning
Shere Ali; but that then our difficulties will be only
beginning; that event would be followed by such a
state of anarchy as would make it almost inevitably
necessary for us to conquer and to keep the country,
as the temptation to Russian interference would
otherwise be irresistible.

He says, that the great evil and misfortune of our
Indian policy has been its variableness and vacilla-
tion;—that it has fluctuated so much under
successive Viceroys (and this partly owing to
changes in the Government at home) as to produce
the appearance of weakness or falsehood.   Shere
Ali is reported to have said, that there was no

trusting to the English, and that he saw there were 1878. only two real powers in the world—the Emperor of Russia and himself!

I have in these last few days read several of the principal speeches in both Houses in the great Debate (which is not yet concluded); those of Lord Derby, Lord Caernarvon, Lord Grey, the Lord Chancellor; Lord Salisbury, Mr. Whitbread. I am not shaken in my conviction that the war is an unnecessary, and therefore unjust one. I think, indeed, it appears from the showing of both sides that we have bullied and ill-used the Ameer, and the great effort on each side seems to be to show that the other party is responsible for this ill-usage. The speeches which I particularly admire for their high moral tone are those of Lord Caernarvon and Lord Grey.

I am very glad to find that William Napier (who has read the blue books), is of our way of thinking on this subject. I have great confidence in the soundness of his judgment and in his thorough rectitude, intellectual as well as moral.

Louis told me that Mr. Whitbread, who led the attack in the House of Commons on this occasion, though a steady Whig, has always been noted as a moderate, temperate, reasonable politician, by no means a regular party man; so that his taking such a prominent part at this time is considered an indication that the moderate Whigs

1878. look on this subject in the same light as the more *advanced Liberals.*

Weather very severe; thermometer in our garden last night down to 15 deg. Faht., or 17 deg. below freezing point. Every night since the beginning of this month it has been below 32 deg.; and ever since the 7th, the ground has been covered with snow; till to-day also there has been a thick frost fog, or rime frost; but this day is very fine, bright, and sunny; the trees, especially the Cedars, look very beautiful in their drapery of frozen snow. There is every appearance of the frost continuing

---

The division in the House of Commons shows a majority of 101 for Ministers—328 to 227. I expected quite a large a majority. Let us hear what the constituencies may say at the next general election.

We were shocked and grieved yesterday by the news of the death of Princess Alice, the Grand Duchess of Hesse. Not surprised indeed, for the last previous accounts of the progress of her malady had been very alarming; but still the case has been thought not to be hopeless. It is a very sad event. What is very remarkable is, that it should have taken place exactly on the anniversary of her 1878. father's, the Prince Consort's death, 17 years ago.— One feels most for the poor Queen, to whom this loss must be a grievous blow, and who has been tried by so many sorrows and anxieties ; but, from all one hears of the Princess, her death must be a severe loss to all who were in any way connected with her, either as relations or friends, or as subjects or dependents. It will be a terrible loss to her numerous young children, although some at least of them are too young to be aware of the disadvantage which they will suffer.

The death of Princess Alice seems to have caused a very general emotion of sorrow and sympathy throughout the country. It is seldom, I should think, that royal personages are so widely and so sincerely mourned as she and her father.

December 19th.

Went through the plant-houses in the garden, to see how their treasures look, after so severe an inroad of frost. Glad to see no mischief done as yet. Every day since the setting in of the severe weather, a blackbird *with a large patch of white feathers on each side of his head* has shown himself on the gravel walk before my study-window, among the other birds which come to feed on the crumbs thrown out. He is a male bird, with a fine orange-coloured beak ; and is very bold.

LETTER.

Barton,
December 24th, 1878.

My dear Edward,

1878.   Fanny and I most heartily wish you a happy Xmas and New Year, and many of them. I do not know where you may be at this present time —I suppose not in London, but in one or other of the friendly houses in which you usually spend your Christmas; but I suppose this letter will be forwarded. Wherever you may be, I hope you have not caught any cold, nor in any way suffered from this long and severe frost. I dislike this " seasonable " weather very much, but I am thankful to say, that Fanny and I have hitherto escaped colds, and are in very tolerable health, and Arthur is very well. Poor Cecil is now with us, and Clement is coming this afternoon, but we shall have no *party* till the middle of January, when there is to be a ball at Bury, and another at Stowlangtoft, and we are going to fill our house for the occasion.

I know of no news in this neighbourhood, nothing in fact but what is in the newspapers. I am reading Burke's writings on the French Revolutions, and reading to Fanny in the evenings, Miss Burney's " Cecilia," which entertains us very much. I hope the printing of your book has been going on prosperously and to your satisfaction. Again wishing you most sincerely all the good wishes of the season,

Ever your affectionate brother,

CHARLES J. F. BUNBURY.

JOURNAL.

A genuine old-fashioned Christmas; weather 1878.
bright and fine, with intense frost,—thermometer
last night down to 11 deg. Fahrenheit; the ground
covered with crisp snow, which has hardly thawed
perceptibly since the 10th or 12th. There is no
appearance of any approaching change.

The news from Afghaunistan continues good,
inasmuch as our armies have made very considerable
progress with little loss. They have now entered
Jellalabad without resistance, and the inhabitants are
said to be friendly. It is believed that Shere Ali
has fled from Caubul, but it is not quite clear that
this is good news, as it may be puzzling to know
what to do next.

Again I feel myself called upon, near the end of
another year, to return my humble and earnest
thanks to Almighty God for the many and great
blessings I am permitted to enjoy. Though very
near the completion of my 70th year, I still enjoy
very tolerable health, and have passed through this
year without any serious or painful illness; and
what I am still more thankful for, I can say the
same of my beloved wife. One beloved and lovely
young kinswomen, Susie Napier, we have lost
within the last two months, but we have good

1878. accounts of Emily (who is at Bournemouth with her mother) ; there seems to be no present ground for uneasiness concerning her ; and dear Minnie is in a better state of health than a year ago. Dear Sarah's safety, and satisfactory recovery from her confinement is a great comfort ; and though there was a considerable alarm a little while ago about her new baby, he seems to be now doing well. The accounts of that most delightful little fellow, Charlie Seymour, continue to be as pleasant as possible. My nephew William (Willie) who entered the artillery service a little while ago, has been very unfortunate; in some sport on board ship, he suffered a strange and severe accident, by which his right hand was so torn, that it is to be feared it will be to some degree crippled for the rest of his life. The last accounts indeed tell us that all danger is over, and that he will not be disabled for the army, but that the thumb of his right hand will probably be permanently useless. A most unfortunate beginning of his military career, poor fellow—to be severely wounded without having been near an enemy.

December 28th.

A welcome change of weather ; the hard frost and snow which has lasted nearly three weeks, broke up two days ago, and even yesterday the snow had almost entirely disappeared, except from very shady and sheltered nooks.

Received a charming letter from dear Rose Kingsley. She tells me that her mother has again been very ill, confined to bed for three weeks;

but that she has rallied very much ever since the 1878.
arrival of her daughter-in-law (Maurice's wife) and
her child; so that the doctor is delighted to find
that the baby has effected more good than all his
prescriptions.

Speaking of Princess Alice in this letter, Rose
says :—" She had always been so faithfully kind to
" my father and to me, that we had a strong sense
" of personal loss in her death. The last time I saw
" her was nearly four years ago, at Oxford; where I
" had a great deal of deeply interesting talk with her
" and thought her one of the most cultivated and
" brilliant, as well as one of the sweetest and most
" womanly women I had ever met."

Excluding near relations, I think we have not lost
any friend in the course of this year, except Sir
Frederick Grey; he indeed is one to be always
regretted. That we have not lost more is rather
remarkable, considering the great age of some of
our friends ; Mrs. Rickards, Mr. Mills, Mr. and
Mrs. Samuel Smith, Mr. Bentham.

The most remarkable persons, not of my actual
acquaintance, who have died in this year are :—
Princess Alice, Pope Pius the Ninth ; Earl Russell,
Lord Chelmsford, Victor Emanuel (the first King of
Italy), Bishop Dupanloup, Mr. Russell Gurney, Sir
George Back, Elias Fries, Padre Secchi, Sir
William Stirling Maxwell, Bishop Selwyn, Sir
Francis Grant, Sir George Gilbert Scott, Sir
Hastings Yelverton, Sir Richard Griffith (the Irish
geologist), George Cruikshank, Mr. G. H. Lewes
(author of a famous " Life of Goethe.")

1878.   Even since the 28th there is another name to be
added to the obituary of this year :—that of Lord
Tweedale, much better known as Lord Walden : a
zealous and distinguished naturalist, and President
of the Zoological Society.   He has survived his
father but a short time.

The year is closing with a gloomy aspect, at
least as to the home affairs of the nation : for I fear
it is certain that there is a very deplorable and
wide-spread state of distress in our great towns and
in all the great centres of manufactures and
commerce.   We must trust that the Almighty may
be pleased to grant us happier prospects for the
coming year. *Laudes Deo.*

1879.

———•◦•———

JOURNAL.

THE New Year was ushered in last night by a violent storm of wind from the S.W. The sound was tremendous, especially about the southern angle of the house ; the outside shutters of my dressing room were torn off and driven against the window, which was shattered. But the storm, judged by its effects, does not seem to have been so severe as the noise led me to suppose. I have, this morning, gone round the pleasure-ground and the arboretum, and can see no damage done beyond the breaking off of small boughs.

———

A startling change of weather ; a return of winter. On looking out this morning we saw the ground covered deep with snow, and it lay all day with hard frost. The snow-fall indeed was greater than any in December.

——— —

With the return of the severe weather, our friend the white-faced blackbird, which I mentioned last month (December 19), has reappeared, coming to pick up the crumbs which I throw out. This

1879. morning I saw a Nuthatch also come for the same
purpose, within a yard of my ground-floor windows.

_____

January 4th.

Our poor widowed nephew, Cecil, who has been
with us since the 23rd of last month, went away
this afternoon.

_____

January 10th.

While there is so much distress in many parts of
England, aggravated by the severity of the season,
it is a comfort to find that in this parish there is no
man out of work, and we hear of no ill-will between
the peasantry and their employers. It is indeed
very good of the farmers to keep on their labourers
while their own condition is so unprosperous, and
while the rigour of the frost and snow makes it so
difficult to find useful employment for them. At
Mildenhall, too, I hear from Mr. Livingstone the
Vicar, as well as from Scott and Betts, that there is
no considerable distress.

The severe season seems to make itself felt far
and wide. I read in the paper of deep snow in
Devonshire, many roads and lanes rendered
impassable by it: and of the railways in Central
France completely blocked by snow, and communications with important towns cut off.

_____

January 12th.

Harry has passed his examination at Cambridge
—and creditably, having qualified for an ordinary

degree, though he has failed to obtain mathematical 1879. honours. This is owing to the difficulty he had in passing the preliminary examination (the "Little Go") in classics, and his repeated failures in it, which left him no sufficient time to "get up" the higher mathematical subjects. His college tutor, Mr. Pattrick, professes to be well satisfied with his success : and indeed I had not, for some time past, expected anything more.

LETTER.

Barton,
January 20th, '79.

My Dear Edward,

A great many deaths seem to have been caused by the severity of the season. Poor Lady Bristol has lost her sister Mrs. Fitzwilliam, who was carried off by a very rapid attack, so that she (Lady Bristol), hurrying off from Ickworth on the news of her danger, could not arrive in time to see her alive.

You will have seen in the papers the death of Henry's friend Colonel Duff, as well as of Mr. Yorke the member for Cambridgeshire.

I should think there has been no winter for a long time attended with such great falls of snow : at least it seems very strange to read of roads and lanes and even railways in *Devonshire* blocked up by the snow ; it reminds me of our father's description of his journey in the winter of 1813-14. It is too

1879. soon yet to judge of the amount of damage which may have been done by the frost and garden and arboretum.

We are both well, I am thankful to say, but are much grieved at hearing such a sad account of Helen Blackett. Your letter to me was the first intimation we had of her illness, and to-day William Napier writes that she is dying. I am very sorry, I have a great liking for her.

With Fanny's best love, believe me,

Your affectionate brother,

CHARLES J. F. BUNBURY.

Have you seen anything of Mr. Spencer Walpole's " History of England," since 1815? The article in the *Edinburgh*, which I have just read, gives a favourable impression of it.

How is your book getting on?—well, I hope.

JOURNAL.

January 22nd.

Our dear boy, Arthur MacMurdo (almost like an adopted son to us), set off yesterday morning on his way to Darmstadt, where he is going to study military science under Col. Wilkinson, as a preparation for Sandhurst. The parting with him is a great effort and a great pang to Fanny who feels quite like a mother to him : but she has wisely and bravely made up her mind to it for his good.

We had a numerous and very merry party in the 1879. house all last week, for the sake of three balls, or dances. First, Mrs. Wilson's ball at Stowlangtoft ; second, a public ball at Bury ; and third, our dance here. Our house party included some very pleasant people :—in particular, our dear Isabel (Hervey) Calliphronas, and her husband, whose acquaintance I was very glad to make. She is natural, frank, good and charming as ever. He is thoughtful, intelligent, and very well read. Lady Alfred Hervey and her very lovely and interesting daughter. Mr. and Mrs. Sancroft Holmes from Gawdy Hall in Norfolk. He is a very intelligent, cultivated, very agreeable young man : has travelled in various countries, even as far as Ceylon, has observed much, and seems to have thought much on various subjects.

Mrs. Holmes (who is a new acquaintance), is very handsome.

January 24th.

We had the satisfaction, yesterday evening, of learning by a telegraphic message that Arthur had arrived safe at Darmstadt.

January 31st.

We had a very agreeable party in the house from the 24th to the 29th ; Lady Mary Egerton, her daughters Emily and Georgina, Mr. and Mrs. John Martineau, and (part of the time) the Barnardistons and John Hervey.

1879. The Egertons and Martineaus are full of mental activity and interest in intellectual subjects, so that I always feel that I profit by their society independently of their agreeableness. Emily Egerton is quite charming : so is Mrs. Martineau.

This month has been one of the most steadily and uniformly cold that I can distinctly remember.

February 1st.

A remarkable event has just now happened in France :—a change of Government without a revolution,—at least without violence.

Marshal MacMahon, finding that his Ministry, as well as both Chambers, were against him on the question of military patronage, has quietly resigned, without a struggle. He has acted like a wise and good man in a constitutional and dignified manner. M. Grévy, who has been elected President in his stead, is supposed to be a moderate Republican— not a Radical, nor inclining to the "Rouges," though free from any leaning to Royalty or Imperialism. I hope he will fulfil the expectations entertained of him, and that France will continue to prosper in peace and quietness.

February 4th.

My birthday, on which I completed my *seventieth* year. I feel that I owe most deep gratitude to Almighty God, for permitting me to live to this age, in good health, and surrounded by so many blessings. It is no small mercy that I find myself at 70, free from any painful or troublesome ailment, and though not strong, yet exempt from any positive infirmity except deafness—and *that* not extreme. It is a still greater blessing that my beloved wife is also in good health, and that we are as firmly united in bonds of love as ever.

I am not conscious as yet of any failure of mental power, or of the capacity of enjoyment either from intellectual activity, or from affection and friendship. Indeed, the loss of friends by death, of which we have of late years had so much experience, the loss of old friends, and, what I feel as still worse, the loss of dear ones of the younger generation—this is the chief cause which casts (as Gibbon expresses it) " a browner shade over the evening of our days."

Susan Horner, in a very agreeable and interesting letter which she wrote to me on the occasion of my birthday, says very well : " I feel as if each decade of our lives were a sort of landing place, from which we look backward and forwards." I feel the same ; I cannot help fancying that every tenth year is somehow a more marked and more distinct step or stage in our progress towards the end, than other years ; and the 70th seems more especially defined.

But the view from it looking forwards (with reference to this life only), seems very much con- tracted, in comparison with the view backward. I often indulge in the retrospect ; not that it is *entirely* pleasant ; far from it ; innumerable faults and follies, small and great, crowd upon my

1879. memory.   But that memory is occupied by the
bright images of so many happy days, so much love
and kindness, so many precious friendships, so much
time passed in agreeable and interesting pursuits,
so much enjoyment derived from the works of the
Creator, that all this is amply sufficient to make me
feel deeply grateful for the past.

LETTER.

Barton,
February 7th, 1879.

My Dear Susan,
          I thank you most heartily for your very
kind letter and good wishes on my 70th birthday:
and I thank you and Joanna very much for the
admirable photograph from Perugino, with which
Fanny is greatly delighted.   I quite feel the truth of
what you so well say, that each decade of our lives
is a sort of stage or landing place from which one
may look backward and forward; and I think the
70th year is a peculiarly marked and impressive
station of this sort; though the forward look must
needs be a very short one.   You judge me very
kindly, and indeed, too favourably.   I cannot say
that I can look back with any great degree of self-
satisfaction; though I certainly do so with an
intense feeling of gratitude for the innumerable
blessings which I have enjoyed in the course of
my life, and which I still enjoy.   It is no small
mercy that I am now at the entrance of my 71st

year, free from any painful, or oppressive, or
harassing disorder, and from any infirmity that I
know of except deafness. I must not complain of
the decline of strength, and indeed, I never was an
athlete, nor ever wished to be. But I cannot help
now and then regretting that I cannot walk well,
which incapacitates me for botanizing, except in the
study, or in my garden. But want of practice may
have something to do with this; and, at any rate,
I have abundant cause for thankfulness for my own
health, and still more for Fanny's.

I most heartily feel the truth of all you say on
this subject, and on the happiness which the
young growing up around us contribute to the
evening of our lives.

To turn to your friend* who is such an admirer of
Shelley, I suppose he knows Trelawney's "Memorials of Shelley and Byron," published not long
ago—a book which has interested me a good deal,
and has given me a much more distinct idea of
Shelley, and at the same time a more favourable
impression of him, than I had before.

I lately read Taine's book on England; what
reminded me of it was, that he speaks of England
as superior at the present day, to all other
European nations, in poetry; at the same time
that he thinks very meanly of our painters. The
work which he selects as the especial type and
glory of our modern poetry is Mrs. Browning's
"Aurora Leigh," which I never read. I dare say

* Captain Silsbee, of Salem, near Boston, Mass.

1879. your friend's explanation of the inferiority of our painters is very true—that they do not study enough ; in fact that the connoisseur's criticism in " The Vicar of Wakefield," is applicable to them.

(Feb. 9th). Here is such a change of weather. that I am now writing in my dressing room without any fire, and it feels quite warm.

The snowdrops and the little yellow aconites have come into blossom almost suddenly, and the air feels like spring. It is a delightful change, and though I dare say we shall have more or less of a return of winter, we may enjoy the fine weather while it lasts.

I leave it to Fanny to tell you of our social pleasures, of which you will have a full account, no doubt, in her journal. I will only say that we have just now a particularly pleasant party, and in particular, dear Kate Hoare with us, which is to me, a special delight ; she seems to grow more and more charming.

We have good accounts of Arthur from Darmstadt, and of Cissy and her daughter from Bournemouth.

With best love and many thanks to dear Joanna,

Believe me ever your loving brother,

CHARLES J. F. BUNBURY.

JOURNAL.

February 11th.

Most deplorable news from South Africa. A

force consisting of part of our 24th regiment and 1879.
600 natives under British officers, has been totally
defeated and almost destroyed by the Zulus; 500
soldiers and 30 British officers killed, a great
quantity of arms and ammunition, and even the
colours of the 24th taken by the enemy. It is a
terrible disaster and disgrace. There are no
particulars yet, nothing to explain how it happened;
but we have suffered nothing near so bad, I should
think, since the Caubul business in '42. The loss
of so many officers is most lamentable. What
could possess us to go out of our way to attack
those wretched savages, and then to conduct so
unskilfully, what we began unjustly.

Yet Sir Bartle Frere is a very good man, and I
had supposed him to be a very wise one; there is
something in the business which requires explanation.

——————

February 13th.

I believe I was hasty in pronouncing that our war
against the Zulus was an unjust one.

══════

LETTER.

Barton, Bury St. Edmund's,
February 13th, 79.

My Dear Katharine,

I thank you very much (though I am
afraid my thanks are a little out of date), for your
kind letter of congratulation on my birthday. I do

1879 indeed feel that I have very great cause for thankfulness for having arrived at this age in such good
health, and in the enjoyment of so many blessings.
It is a great mercy that Fanny as well as myself is
in good health, and that we have passed so far
through this severe winter without any illness—in
my case—without even a common cold. But we
must not "halloo till we are out of the wood," and I
suspect we are yet far from the end of *the wood*—
(winter, that is)—so I think you are quite right in
recommending caution and prudence.

I was very sorry to hear you had such a severe
and painful rheumatic attack : but I hope you have
by this time quite got over it.

What a dreadful, deplorable disaster this in South
Africa! I do not think that anything so bad (in a
military way), has happened to us since 1842. A
disgrace in a military sense : and disgraceful to the
Government if (as seems to be thought), they
neglected to send out sufficient forces to support
the defiance which they had given to the savage
king. Cetewayo seems to be just of the same
stamp as the Zulu chiefs, his predecessors Chaka
and Dingaan, an African Attila or Jenghiz Khan.
No doubt, in the long run we shall conquer the
Zulus, and (I hope) improve Cetewayo off the face
of the earth; but in the meantime, how many hearts
are sorrowing for our brave men who have fallen !
I am afraid the Godwin-Austens have lost a son in
this melancholy business.

I have nearly finished reading Hookers' and Ball's
"Morocco," and have read it with great pleasure,

being much interested in the Botany, of which there 1879. would be rather too much (I should think) for "the general reader." I wanted particularly to know whether they found any of the remarkable plants of Madeira and the Canaries, which have hitherto been found nowhere else : and it is very striking and curious to find that no trace of them was discovered in Morocco—except indeed some fleshy Euphorbias differing only specifically from the Canary one. In other respects the country does not seem to be a particularly interesting one ; and as to its political and social relations, Cetewayo and his Zulus can hardly be worse. All our old notions as to the pre-eminent barbarism of Morocco and tyranny of its sovereigns are confirmed or surpassed.

During the cold weather which so long confined me to the house, I got on little by little with my Botanical Notes ; also made a list of our cultivated Ferns, and began one of my collections of Mosses. I am still reading Miss Burney's "Cecilia" to Fanny when we are alone in the evenings, for we have been so often interrupted by company that we have been a long time about it ; and indeed, though there is much in it that is interesting and entertaining, it is certainly a long-winded book.

Have you read the article on Bismarck in the last number of *The Quarterly?* It is extremely good, I think ; very severe and not less just.

In *The Edinburgh* I was interested by the article on Spencer Walpole's "History of England," since 1815, and that on Mrs. Jameson.

1879.   Our last party (of which Fanny will have given
you an account), was a particularly pleasant one
to us.

Pray give my love to dear Rosamond, and
believe me,

Ever your affectionate brother,

CHARLES J. F. BUNBURY.

JOURNAL.

February 15th.

A particularly pleasant social party* in our house
last week and the beginning of this; from the 6th to
the 12th:— Dear Kate Hoare and her husband,
Leopold and Lady Mary Powys, Colonel and Lady
Constance Barne, Colonel and Mrs. Corry, Mr. and
Mrs. Montgomerie, Miss Loch and John Hervey.
The Wilsons and some others of our neighbours
also dining with us.

Mrs. Corry, daughter of Lord Halifax, is a very
interesting agreeable person: very intelligent and
well read, and of very attractive manners. I think
she will be a great acquisition to our acquaintance.

February 25th

We are still without official or thoroughly
authentic accounts of the terrible slaughter of our
troops on the 25th January; and the reason appears
to be, that scarcely anybody escaped from it.

It was at this party that we promised to give a new billiard table, as the old
one was worn out.—F. J. B.

There is, however, in *The Daily News* of yester- 1879. day, a rather full and intensely interesting narrative by a Captain Young, who seems to have been, by some chance, a mere spectator of the early part of the fight, and to have escaped at last with great difficulty when nearly surrounded by the Zulus. There is a good deal however, as yet, which is hard to understand; and one cannot yet feel quite satisfied of the authenticity of the narrative. It is, however, probable enough in itself. The Zulus seem to have not only had an enormous superiority of numbers, but to have fought with astonishing courage and determination, and even with a good deal of military skill.

It is evident that they have now become very formidable enemies. It is said that they attacked in regular divisions, in compact order, eight deep, with skirmishers thrown out in advance: that they began with a fire of musketry, and when they had advanced near enough, threw away their muskets, and rushed on to close fight, stabbing with their short spears, and engaging so closely that our men had scarcely room to use even their bayonets. They must be very different from our old antagonists the Kosa Caffers. There is a great deal that requires explanation, as to the steps which led to this disaster—how the column came to be so isolated, and so forth; and part, I suppose, will never be cleared up. There has since it seems, been a partial engagement, in which Colonel Wood has repulsed the Zulus; and by the latest telegrams it appears — at least we may hope, that our

1879. remaining forces on the frontier occupy defensible
positions.

Our poor friends the Godwin-Austens have lost
one of their sons in this terrible fight. They had
two sons in the 24th, but it seems that one was not
in the battle.

February 26th.

Mrs. Clements Markham showed me an interesting little collection of Arctic plants, given her by
one of the officers who went on the last Arctic
expedition. They are all (as one might expect)
plants of very humble growth, but several of them
very pretty. All were known to me by descriptions,
from books, but of many I had never seen
specimens. Those which interested me most
were :—

> Saxifraga flagellaris.
> Saxifraga oppositifolia.
> Saxifraga cæspitosa.
> Saxifraga cernua.
> Draba alpina.
> Epilobium latifolium.
> Papaver nudicaule.

March 1st.

There is now some additional news from South
Africa—a week later than the last we had before.
No further disaster, and indeed no event of im-
portance, has since happened. Our forces are

wisely, acting strictly on the defensive, guarding the 1879.
frontier of Natal. The colours of the 24th regiment
have been recovered, which is satisfactory. They
were found lying on *our* side of the river, beside the
bodies of two officers, Mr. Coghill and Mr. Melville,
who were mounted, and who seem, though mortally
wounded themselves, to have carried them out of
the fight. Brave men. There is nothing yet that I
see, to explain clearly how the disaster happened.

Yesterday and to-day were my Rent Audit days
at Mildenhall, and the result is very satisfactory.

Scott writes to me:—

"Although the tenants, like all farmers, are
" feeling the effects of two or three bad seasons in
" succession, with unusally low prices this year, the
" same excellent feeling exists to its full extent: not
" a disagreeable word was uttered by one of them,
" and all appeared to have perfect confidence that
" they should always meet with just and liberal
" treatment. This is the more satisfactory from the
" fact, that the estate adjoins (or nearly so), *six*
" parishes, which at the present time are said to
" contain *between them*, only four or five tenants, the
" lands being held by the owners themselves, the
" occupiers having left them." The rent which he
has collected for me this time, is fully equal to that
which has been usual.

I must say that much of this good feeling on the
part of my tenants, which is so gratifying, is really
to be ascribed to the excellent manner in which
Scott has managed the estate—to his moderation,
good sense, good feeling and right principle.

Remarkably agreeable company here, since the
18th February: Mr. and Lady Muriel Boyle, Captain
and Mrs. Douglas Galton and their daughter Gwen-
dolen, Mr. and Mrs. Walrond, Mr. and Mrs. Clements Markham, Captain Markham, Edward Campbell
and his daughters Annie and Griselda, Lord John
Hervey. Mr. Walrond, Mr. Courtenay Boyle, and
the two Markhams, combine social pleasantness
in a remarkable degree with intellectual vigour
and varied knowledge. Douglas Galton is certainly
a man of great ability. Edward Campbell is one
of the most amiable, of the kindest, best-hearted,
and altogether most excellent men that I know;
has seen much service, and known a variety of
men, and is very pleasant when not oppressed by
low spirits, for he has never entirely rallied from the
death of his delightful wife, though it is six years
since.

As for the ladies of our party—I have already
often mentioned and praised Lady Muriel Boyle,
and I like her, if anything better than ever; she
is quite charming. Mrs. Markham is very clever and
well-informed, with lively, pleasant manners.

Annie Campbell, a delightful girl, very pretty,
with manners both lively and engaging, and one of
the most admirable dispositions and characters in
the world. It is beautiful to see the devoted
affection between her and her father. Her little
sister Griselda is a most beautiful child.

———

Lord Chelmsford's dispatch concerning the disaster at Isandula has at last arrived and been published; but it does not yet seem to make the affair quite clear; in fact, he himself says that the blow appears to him almost incomprehensible. I am afraid it must have been owing to rashness, that the unfortunate 24th left their camp to attack or to pursue the enemy, and were surrounded by their overwhelming numbers. It will not be the first time that our troops have suffered disasters owing to a rash contempt of barbarous or savage enemies; it is said too, that on this occasion they neglected to fortify their camp, as they might easily have done with waggons fastened together after the manner of the Boers. Edward Campbell thinks it strange that Lord Chelmsford did not attend to the fortifying of this camp before he advanced with the other column.

The successful defence of the position at Rorke's Drift, by two officers and a mere handful of soldiers, against a great mass of Zulus is very fine, and redeems the honour of our army. The names of the two officers—Chard and Bromhead—ought never to be forgotten.

March 6th.

The Times correspondent, in to-day's paper, gives a very long and very interesting description of the disaster at Isandula. He was with Lord Chelmsford's column, therefore was not an actual eye witness of the scene; but from interrogating the

1879. few survivors, he gives us what (put together with
the other fragments of narrative) is probably the
best account we shall ever have of the melancholy
business.

It seems clear that the Zulus completely out-
manœuvred our troops, surrounded them with an
immensely superior force, and then attacked at
close quarters with astonishing courage and resolution.

It appears also, that Lord Chelmsford with his
column, narrowly escaped running into the very
jaws of the victorious enemy.

March 21st.

Our poor friend Lady Grey (Sir Frederick's
widow), left us yesterday, having been here since
the 12th. I noted in my journal of May last year,
the death of that excellent, very able, and very
agreeable man, her husband. She is very calm
and composed, at times even cheerful in conversation; but the change in her looks sufficiently
shows how much sorrow she has suffered. She is
very, very lonely, having not only no child, but no
sister living. I never saw a married couple who
appeared to me more thoroughly united, more
absolutely all in all to each other than she and
her husband, and the desolation must be in
proportion. She has found it advisable to sell
the beautiful little place they had—Lynwood, near
Sunningdale, on the edge of Windsor Park (a
perfect gem of a place it was) and to buy a smaller

one near Weybridge. Lady Grey had a niece with 1879.
her while here, Miss Sullivan,* one of her brother's
daughters, a very nice girl.

March 22nd.

I am very sorry for Sir Bartle Frere. I have a
great liking, and a great respect for him, and I feel
sad that so late in his life, after so long and so
honorable a career, he should have come to incur
so much blame and reproach.

I have no doubt that in all this unhappy Zulu
business, he has acted strictly according to what he
thought right; and I am quite aware that his
position was a very difficult one. In the last
dispatches which passed between him and Sir
Michael Hicks-Beach (I judge from those portions
of the correspondence which have just been
published in *The Times*), he has expounded his own
views with great clearness and ability ; but I cannot
help coming to the conclusion that Sir Michael is in
the right.

Sir Bartle appears to me to have rushed into the
war hastily, and with unwise precipitation, contrary
to the clearly-expressed opinion, and almost to the
instructions of his Government. He is now in a
very disagreeable predicament, for the opposition
are bitter against him, and it seems that the Government will make but a luke-warm defence.

March 27th.

The Ministry had a great majority on the question

---

* A very charming girl. She died at Fairmile House, Cobham (her Aunt's),
1889.—F. J. B.

1879. about Sir Bartle Frere, more than two to one—as
was to be expected, since it had been turned into a
motion for a censure on the Government, and so had
been made a decided party question; as such
therefore the Members voted on it.

Accordingly the Ministerial speakers all took Sir
Bartle under their protection; and Lord Cranbrook
in particular, did all he could to explain away and
nullify the strong dispatches in which Sir Michael
Hicks-Beach had blamed him.

The speeches, of which I read the full reports
in *The Times*, were, on the one side,—Lord Lands-
downe, Lord Blachford, Lord Kimberley.   On the
other—Lord Cranbrook, Lord Carnarvon, Lord
Beaconsfield.

April 1st.

Ministers have had a much less majority in
proportion in the House of Commons than in the
Lords; a majority of 60.   Such as it is, it cannot
be considered as a real judgment on the merits of
the cause; it is a party verdict, for no doubt all
voted against the censure who thought it undesirable
that the Ministry should go out.

April 3rd.

We have lately had, in a letter from Susan Mac-
Murdo to Fanny, the sad news of the death of the
MacMurdo's son-in-law, Mr. Johnstone, the husband
of their daughter Susan.   He died of inflammation of
the lungs, at San Remo; had apparently not been

long ill. I did not myself know him, but the Mac-Murdos seemed to have a regard and esteem for him ; and it is very melancholy that his poor young wife (only 20 years old), should be left a widow after so short a period of wedded happiness and without a child. He had succeeded to a handsome fortune only a short time before his death, and she is left rich ; exempt thereby from many trials and anxieties, but exposed to temptations of another sort. She is a beautiful creature.

April 7th.

Katharine lent us, yesterday, a correspondence (printed, but I think not yet published), between Sir Bartle Frere and Bishop Colenso, relating to the causes and origin of the Zulu war. They are very civil to each other, but are greatly at variance in their opinions on the subject—the Bishop urging strong arguments against Sir Bartle's Ultimatum, and particularly as to his treatment of the Zulus on the question of the disputed territory. In a private letter to one of his own family, written after the declaration of hostilities, he calls the war an *unjust* and *unnecessary* one. The Bishop may be partial—probably is ; but at any rate he knows the people much better than Sir Bartle can. I repeat I am sorry for Sir Bartle Frere ; personally I like him much, and I have no doubt he is a good man as well as a very able one, but I am afraid the Bishop is right as to the war.

I am surprised however to observe, that the Bishop expresses his assent to the demand (in the

1879. Ultimatum) for the disbandment of the Zulu army, which I should have thought the most offensive demand of all.

News of another disaster in Zululand. A force of 70 men of the 80th Regiment escorting a convoy of waggons, was surprised by a large body of Zulus on the 12th of March, and cut to pieces; Captain Moriarty and 40 men are known to have been killed, and 20 were missing. Certainly we have either very bad luck or bad management in this war.

Mr. Clements Markham left us this morning, having been here since the 7th, engaged in making extracts from some of the rare old books (collected by my father) on military science. He is engaged in writing a history of the great family of De Vere.

Clements Markham is a man who has a great deal of conversation, and is well worth listening to; he is very clever, lively and animated, has a prodigious amount and variety of knowledge, and is very ready to communicate it. The range of his information is indeed astonishing. Besides being one of the greatest authorities in this country—if not the greatest—on Geographical science, (he is Secretary of the Geographical Society), he is devoted to Bibliography, and seems, as far as I can judge, to be very learned in old books and all

relating to them ; very great also in genealogies,
and acquainted with the history of innumerable
families, and moreover, he has an extensive
acquaintance with men as well as books, and with
the recent as well as the old history of families. In
short, he is a very agreeable as well as learned
man, and one from whom one can learn a great
deal. I wish I could remember any tolerable proportion of what he said.

Talking of the English habit of mangling foreign
names, Mr. Markham said he had seen an English
pamphlet on Indian affairs, of the last century, in
which the name of the cruel Nabob of Bengal was
printed, "*Sir Roger Dowler.*"

Mr. Markham observed that most of the provinces
of France, under the Ancien Régime, had had
provincial Parliaments, or other local institutions,
which—though they might have been reduced to a
weak and dependent state under the later Monarchy
—might have afforded a valuable framework for a
free Government: and that it would have been far
wiser and better to build on those "old lines"
instead of demolishing everything before they began
to construct. I quoted Burke's description of the
Revolutionists as mighty *architects of ruin* : and he
admitted its justice.

Mr. Markham observed that Motley, in his
History—especially in the latter volumes—was very
unfair to the English, and showed a strong
prejudice against them, particularly against the
Veres. This was probably owing much to his
Republican zeal, which was inflamed by the Civil

1879. war in America. The Spanish historians, on the contrary (Mr. Markham says), show a remarkable fairness in relating the history of those times, and do justice to the courage and skill of the English and even of the Dutch.

April 14th

Very extraordinary weather for the middle of April. The wind on the 11th was very cold and rough ; in the night of the 11th, or early morning of the 12th, the thermometer in my garden went down to 22 deg. : hail began to fall soon after noon on the 12th, and soon changed to snow, and a regular heavy fall of snow set in, and continued steadily till dark ; whether it went on all night, I do not know, but it was still snowing when I looked out in the morning of the 13th. It ceased, however, before noon, and the sun came out bright. It has been a much heavier fall of snow than any in this winter. Scott found the depth of snow in the morning of the 13th in the arboretum on the level ground, where there had been no drifting, to be fully one foot. This morning, Allan, my gardener tells me that several trees, one or two rather favourites of mine, have been broken down and shattered by the weight of the snow. We have had no such snowfall for some years.

The old Douglas Fir (planted by my father in 1831), has lost so many branches in this snow, that it has a deplorably mutilated look. Our finest Judas tree near the house has lost a great limb. A large and fine Bird Cherry, near the edge of the

pleasure-ground, is completely broken down. These 1879.
are the greatest losses that I have yet seen.
Happily, neither the great Cephalonian Fir, nor my
favourite Catalpa, has suffered.

April 15th.

The shocking news of an attempt to assassinate
the Emperor of Russia. It is a detestable wickedness.

April 17th.

Very much shocked and grieved by the news
(which we saw in *The Times*) that Captain Ronald
Campbell of the Coldstream Guards is one of
the officers killed in Colonel Wood's last engagement with the Kaffers. He is a brother of our dear
and charming young friend Lady Muriel Boyle.
I grieve very much for her ; she loved her brother so
much,—she talked much of him the last time she
was here, and thought him fortunate in being with
Colonel Wood. It is very grievous ; and how many
valuable lives will yet be sacrificed before this
odious, detestable war is over ! What oceans of
blood will be shed, how many homes will be made
desolate !—and all for no advantage to England,
still less for the promotion of any principle. I suppose we shall in the long run succeed in crushing
these fierce and brave savages ; but what then ?
The annexation of all Africa, even to the frontiers of
Egypt or of Algeria, would not be worth the cost
of so much British blood.

George Betts, who came over from Mildenhall,
gave a deplorable account of the prospects of the
crops in the Fen country. On the so-called "high
lands" (those which have a dry soil, and are above
the level of the fens and above the reach of inundation), things are not so bad ; but in the Fens the
young corn, both wheat and barley, have suffered so
much from the long-continued frosts, that it is
doubtful whether they can ever recover. The
turnips also have been rotted by the frost, and it
is difficult to find feed for the sheep. Altogether, it
is a lamentable prospect.

Scott says he hardly remembers to have seen the
young corn look so ill at this time of the spring as it
does now.

News of more fighting, and very severe, in South
Africa ; this time ending in a victory for us, with
great slaughter of the enemy, but also with heavy
loss to us. Lord Chelmsford seems to have managed the relief of Col. Pearson well and cleverly ;
but it becomes more and more evident that the
Zulus are fierce and formidable enemies. This last
engagement was on the 2nd (if I understand rightly),
of this month.

Col. Wood's two actions seem to have been on
the 28th and 29th of March, but the accounts of

them are still very confused and hard to make out. 1879.
It was in one of these two days that poor Captain
Campbell was killed.

Col. Nason remarks that the peculiar mode of
fighting of the Zulus which has proved so formidable,
seems to make it necessary for our troops, when
opposed to *them*, to alter their tactics, to abandon
the system of fighting in loose order, which has
been adopted in our army for the last few years,
and to return to the close order used in our former
wars. The loose or open order has advantages
when opposed to a powerful artillery, but the furious
rush of the Zulus (like that of the Highlanders in
former days), seems to require a compact array and
concentrated fire to meet it.

May 3rd.

Weather, for a week past or nearly so, very fine,
bright and sunny, but constantly cold ; frost nearly
every morning, thermometer sometimes as low as
25 deg. Very bad for the farms ; everything at a
standstill, nothing growing. Few men here seem
to remember so backward a season. Few trees
yet well in leaf. There are great fears for the
harvest, and another very bad harvest would be
ruinous.

On the 1st, there was a smart shower of sleet,—
in some places, I have heard, actual snow.

May 5th.

Whatever may be the case with the fruit trees,

1879. this bright, sharp, dry weather, does not seem hurtful to the flower garden. All the periodical phenomena, indeed (flowering and leafing), are from three to four weeks later than last year; but the " spring beds " of tulips, hyacinths, forget-me-nots, arabis, and others, are in full beauty; and the bees —honey and humble bees—resort to them as busily as ever.

May 8th

Mr. Bradshaw, the Librarian of the University of Cambridge, came to us (introduced by my nephew Harry) on the 5th, and stayed with us till yesterday afternoon, when he and Harry went back to Cambridge. He spent much time here in our library, and seemed much pleased with it; much struck and interested by the number of rare and curious books which my father had collected; several of them, he said, were not in the Cambridge University Library. He is evidently a man of much general intelligence, as well as learned in bibliography; and his conversation is good.

May 21st.

Dear Minnie left us, and returned to town on the 13th, and we have been very quiet, only gradually making preparations for *our* move to London.

The weather has at last turned really fine and warm, and the last few days have been delightful, the grass brilliant, the gardens in their glory, the trees clothed with the various exquisitely delicate

colours of their spring foliage, the soft tawny reds of 1879.
the Indian horse-chesnut, the white American
maple, and the purple beech, blending exquisitely
with the tender green of the common beech and
others of the ordinary trees of our country. This
year, the common ash and oak are coming out into
leaf and flower, both about the same time, which
is unusual.

May 22nd.

The newspapers announce that Yakoob Khan,
the Afghan, has accepted the conditions proposed by
our Government as a basis of negotiation, and that
the Afghan war is in effect at an end. I am very
glad of this, but I think we may well ask, why it was
begun, and what has been really gained by it.

May 25th.

Yesterday was a beautiful and delicious day,
coming between two very wet ones.

We drove to Mildenhall, and spent all the afternoon there ; enjoyed very much the beauty of the
country as seen in its spring dress of fresh young
leaves and orchard blossoms. Visited Mr. and Mrs.
Livingstone in their new house in which they now
seem comfortably established; it is now secured
(till the disestablishment of the Church !) as the
permanent Vicarage-house ;— a thing which has
long been a great want in the parish. The final
settlement of this matter has cost me a huge sum of

1879. money, and a still worse heap of trouble and worry. Law business in general is a great plague, but of all lawyers (as far as my ex perience goes) ecclesiastical lawyers are the most tiresome and vexatious — but I am very much compensated for it by finding that I have secured for the parish of Mildenhall such a Vicar as Mr. Livingstone, a highly-cultivated, enlightened, liberal-minded man, a genial and agreeable member of society; what is more important, an excellent and devoted clergyman; and what in these days is more rare, a man of genuinely liberal mind, free from narrowness and bigotry, and concilia- tory in all his intercourse with those of other sects. He is a most valuable man.

Visited also my old friend, Miss Bucke, with whom I used to play bagatelle at Mildenhall, when I was a child. She is a good old soul.

Another valuable man whom we saw at Milden- hall, was Mr. Saxton, the schoolmaster; he, indeed has been established there for very many years, and we have long known his worth.

———

May 26th.

Important news, both from the east and the south :— the treaty with Yakoob Khan actually signed,* and he acknowledged as Sovereign of Afghanistan. Sir Garnet Wolseley appointed both military and political Chief of the districts of South Africa which are the seat of war or of prob- able seat of war. Sir Bartle Frere's authority

* The Afghan quarrel thus set at rest—till next time—as the children say.

being restricted to the old Colony.   This I should   1879.
think is well ; the seat of war is so distant from Cape
Town and the old provinces, that it is too much for
one Governor to attend properly to both ; and
where a country is placed in such difficult circumstances as Natal and the Transvaal, it seems best
that the political and military authority should be
concentrated in the hands of one.

                                                    May 27th.

Up to London, to 48, Eaton Place.

                                                    May 28th.

We went to a large afternoon party at the
Hortons, in Grosvenor Place ; met a great many
friends and aquaintances.

                                                    May 30th.

The 35th anniversary of our happy marriage.   I
feel deeply grateful to Almighty God for the many
years of domestic peace and happiness, and the
many and various blessings which He has granted
to me through this union ; and deeply grateful also
to the admirable and beloved wife through whom so
much good has been bestowed upon me.   The married
portion of my life has now been as long as the
unmarried.   I was 35 years old when I wedded ;
and it is 35 years since.

May 31st

1879.   We had the news of Frank Lyell's engagement to
be married to Miss Guise, daughter of a brother of
Sir William Guise.

Felix faustumque sit!

———

June 1st.

Sir William Guise, whose niece Frank Lyell is
going to marry, is a man with whom I have been
acquainted a long time, but never intimately. I
first met him at Fort Armstrong on the Kat River,
in S. Africa, in 1838, when he was an officer in the
75th regt., and I was travelling with Sir George
Napier in the Cape Colony. I did not see him
again till he was a man of large estate and of
importance in his county. He is a zealous
naturalist, and in particular a learned geologist, has
travelled a great deal, and is an able and well-
informed man. He spent a day or two with us at
Barton in 1868, after the British Association
meeting at Norwich.

———

June 2nd.

Fanny came back from her visits to-day quite
delighted with the account she has heard of Miss
Guise.

It is said, Lord Beaconsfield has remarked that
the Zulus must be indeed a formidable people :—
they have beaten our General, and they converted
our Bishop. (Fanny heard this from Lord
Hanmer).

A fine day—almost the first since we came to London.

We drove round Battersea Park, and enjoyed the beauty of the shrubberries, where there are at last some signs of spring :—the lovely, fresh, tender green of the young leaves, the blossoms of the lilacs, laburnums, thorns, horse-chesnuts, guelder roses, red horse-chesnuts and others in great beauty. But here as at Barton, most things are nearly a month later than in ordinary seasons.

June 5th.

We went to see the Water Colour Institute Gallery (in Pall Mall), and were very much pleased with it. Such a number of beautiful landscapes, smiling woodlands and corn-fields and meadow scenes and wild picturesque subjects of mountain and glen and lake in Wales and Scotland, give one great enjoyment, especially amidst the brick and mortar and rain and fog of London. It is a real pleasure to dwell on such pictures as some of those in this Exhibition of Edmund Warren.— *C. Vacher*, *Mole*, *Hargitt* and a few others. There are also some exquisite little flower-paintings by *Mrs. Duffield*, and Flemish interiors by *Haghe*.

Dear Rose Kingsley arrived — unfortunately to stay only two days. Minnie Napier and Agnes Wilson also dined with us. A delightful evening.

June 6th.

We had a numerous and pleasant luncheon party:—

Lady Winchelsea and her daughter Lady Evelyn Finch Hatton, Lady Rayleigh, Lady Octavia Legge, Sir Lambton and Lady Loraine, Jane Broke, Captain Legge, Lionel Fletcher, Cissy and Emily, Susan and Mimi MacMurdo, my nephew Harry. We dined with the Douglas Galtons. I sat by Mrs. Coore (one of Lord Belper's daughters), a remarkably interesting and engaging person: attractive in face, manners and conversation. Other members of this dinner party were—Dean Vaughan (Master of the Temple), and Mrs. Vaughan, Lord and Lady Hammond, Lord William Paulet, Lady Strangford, Mrs. Ford and Mr. Coore, besides Rose and ourselves.

June 7th.

Dear Rose Kingsley went away, to my great regret. She is going immediately to the Continent, with her sister and brother-in-law, Mr. and Mrs. Harrison.

We went to Veitch's nursery garden, and spent some time there ; saw many beautiful things, but nothing that struck me as very new or remarkable.

June 10th.

Went to luncheon with Miss North, and to see part of her collection of paintings done in India.

The time only allowed of my seeing a part of them.
I have written largely in my journal of '77, of
her exquisite flower-paintings, executed in Brazil,
Java, Borneo, Madeira, Teneriffe, and some other
countries, which I saw in that year. Those she has
now brought home (from Continental India) are to
some extent in a different style from the previous
ones, at least belonging to a different class of
subjects—comparatively few botanical; the greater
part views of scenery and buildings in various parts
of India. They are beautiful and very interesting;
the flowers and fruit as admirable as those of her
former collections; the buildings also; but in some
of the landscapes, I do not know whether an
artist might not object to something of hardness.
I speak however with diffidence and hesitation.

The MacMurdos, to whom Indian scenes were
more familiar than to me, were delighted.

Montague remarked the truth with which she had
rendered the brilliant effect of the white marble in
the buildings, free from the opaque, chalky look
which painting is apt to give. Many of the scenes
are in Rajpootana, a country little visited by
Europeans, and little known to them, as Miss
North said; some of these are very interesting,
particularly those of Odipore and its beautiful lake.
Some of those in the Himalaya, too are very grand.
There are fine portraits and groups of Deodars and
of Abies Smithiana. Miss North's travels in India
must have been very extensive: for she has views
in the Eastern and the Western Himalaya, in the
North-Western Provinces, at Delhi and Agra, at

1879. Bombay, on the Neilgherries, in Malabar and in Tanjore.

Of the botanical subjects here, one of those which struck me most was the Lagerstroemia, a magnificent flowery tree of which I had often read, but had only a very poor idea : and it was here with its curious fruit as well as its beautiful flowers.

One painting with which I was especially struck and delighted, was a view of a Fern-jungle near Darjeeling: the character of the Tree-Ferns rendered with marvellous truth and beauty.

June 12th.

I see in *The Times* of this morning the death of Canon Beadon, in the 102nd year of his age. I mentioned him in my journal when we saw him at the Palace at Wells in 1870 and '71, when he was (at 93), full of life and animation, perfectly clear in mind, and delighted to tell old stories, and talk about Dr. Wollaston and other celebrities he had known.

The last time we were at Wells, we did not see him, as increasing infirmity had made him give up his term of residence. The remainder of his life, I believe was spent quietly in his rectory near Southampton ; and the end of it was very peaceful. He was a good, kindly, genial man.

June 16th.

We went to see the Grosvenor Gallery, but did

not see it very thoroughly. It struck me that there
are not so many fantastic and extravagant pictures
as in the first of these exhibitions which I saw two
years ago. I admired a portrait by *Watts*, of a
dark man in a black velvet dress and a broad black
hat, like a Titian. Also a simple picture of a girl
in a rustic dress, by *Richmond;* " The End of the
Story," by the same ; a portrait of Miss Ada White
Thompson, also by *Richmond ;* and an impressive
portrait of Gladstone, by *Watts.*

June 18th.

Drove in Hyde Park, and the day being very fine
(a delightful rarity, hitherto, this season), enjoyed
the beauty of the young foliage (not even yet soiled
by smoke), and the lovely blossoms of the rhododendrons, azaleas, crimson hawthorns, red horse-
chesnut, and other ornamental trees. The way in
which the park has of late years been decorated,
every summer with flower beds, and with semi-
tropical plants of fine foliage transplanted into it,
adds exceedingly to its beauty. The weather for
the most part, since the beginning of this month,
has been mild, even warm, but very wet : scarcely a
whole day without rain ; but these last two days
have been beautiful.

June 19th.

We went to see Waterer's show of rhododendrons, in Cadogan Place. Very beautiful, like other
shows of a like kind which I have seen in former

1879. years; but it is of rhododendrons alone, without any azaleas or kalmias.

June 20th.

There is a piece of news in this morning's papers, calculated to make one thoughtful and sad. The Prince Imperial—the son of Napoleon the Third—has been killed by an ambuscade of Zulus, in this wretched war. I grieve for his poor mother, so devoted to him, and now so desolate. He seems to have been a fine and interesting young man, and there is something solemnly impressive in the thought of a career so uncertain in its future, yet so likely to be eventful and important, now so suddenly cut short. He was only in his 24th year. It is curious that the only child of the First Emperor Napoleon, and now the only child of the Second Emperor Napoleon, should both have died in early youth.

Dining with Mr. Gambier Parry, I was made acquainted with old Dr. Henry, a very intelligent, scientific old gentleman, who had long been well acquainted with Mr. Horner, and with most of the eminent members of the Geological Society. He has travelled much and his conversation is interesting.

June 24th.

There seems to be a general good and sympathetic feeling in England on the death of the Prince Imperial; an almost universal feeling,

especially for the poor mother. Her situation and 1879.
history are indeed most tragical : quite such as a
Greek tragic poet would have chosen for a subject.
MacMurdo thinks that there was culpable neglect of
the Prince's safety on the part of our General :—
that he ought not to have been allowed to go on
such a service with so small an escort. But he
seems to have been a fine, high-spirited, gallant
young man, and one very likely to run, of his own
accord, into unnecessary danger.

———— ————

June 25th.

At dinner at Lord Talbot de Malahide's, met
Mr. Ball, the botanist, and had some talk with him
about Morocco. He confirmed what I had gathered
from his and Hooker's book, that they did not find
in that country any of the remarkable plants
characteristic of the Canaries and Madeira ; in
particular, none of the gigantic laurels which are
such striking features in the vegetation of those
islands. He is preparing—for *The Geographical
Society's Proceedings*, a paper on the "Vegetation of
Mountains," in which he will bring out some views
which he thinks new and important.

There has seldom been within my memory a
month of June as wet as this one. The rains have
been almost daily ; very few days indeed have been
fine and sunny throughout.

—— ————

June 27th.

The news, that the Sultan has deposed the

1879. Khedive of Egypt at the desire of the English and
French Governments, and that the said Khedive
has submitted quietly.

Curious, to remember 1840, when we thought it
necessary to send a fleet to protect the Turkish
Empire against the Egyptians.

———— —

June 30th.

Our dear Arthur MacMurdo arrived from Darmstadt, safe and well, much grown, looking well, and
in excellent spirits.

Scott writes to me from Barton :—

" Fine weather is much wanted, not only for the
" corn crops, but to secure stover and hay, which is
" now ready to cut. I am afraid dry weather
" coming now would be too late to save the corn
" crops on *strong clay* lands, as the ear of corn,
" being already formed in a small weak state, can
" never become fine and large. I have noticed
" many fields of barley in the neighbourhood,
" including several in this parish, almost spoiled by
" continuous cold, wet weather. Harvest will be
" very late. I have not seen an ear of corn at
" present, and it is calculated that wheat generally
" takes six weeks from the time of its coming into
" ear before ripening, unless very exceptional
" weather takes place."

" In the garden, everything is very backward,
" except the weeds."

" Barton generally supplies a good many young
" potatoes for Bury market early in the season, but

" up to the present time, scarcely a new potato is to 1879.
" be seen. The tops grow rapidly, but the roots do
" not grow, at present, no doubt owing to the
" wetness of the soil. The nights continue warm,
" which helps matters very much ; and many of the
" wheat fields on " *kind* " land continue to look
" well."

July 2nd.

We went to the National Gallery and spent
about an hour there; in the department of British
pictures, with great enjoyment.

It was pleasant to see many poor-looking people
also appearing to enjoy the pictures. Besides our
old favourites *(Callcott, Constable, Eastlake, Gains-
borough, Hogarth, Landseer*, and others), we admired
Crome's singular but strikingly natural Norfolk
scenes, *Thomson* of Duddingston's beautiful Highland
view ; *James Ward's* " Gordale Scar."

Yesterday we had a particularly agreeable dinner
party :—the Bishop of Bath and Wells, Lady
Winchelsea, Sir Francis and Miss Doyle, the Albert
Seymours, the MacMurdos, Sir John Shelley, Lord
and Lady Hanmer and others, nineteen in all. It
was an especial pleasure to have our dear Arthur
Hervey in our house, whom we now so seldom have
an opportunity of meeting. He was looking well,
and was as delightful as ever. Sir Francis Doyle in
great force, very entertaining. Lady Winchelsea,
as I have mentioned before, is a very agreeable and
interesting person.

1879.   July 4th.

We went to see Miss North's Indian paintings,
which are now exhibited at No. 9, Conduit Street;
we spent some time among them with great
enjoyment. The pictures of flowers and fruit are
most beautiful; as I noted before, I think them
decidedly superior to the views of scenery, though
many of these are very interesting. Some of the
botanical subjects which pleased me most (beside
those I noted in '77) were: the flowers, fruit, leaves,
and a reduced general view of the Terminalia,
an Indian tree often mentioned (especially by J. B.
Hooker), but of which I never before could get a
precise idea.

Flowering spike of the Pandanus. The beautiful fruit and seeds of a Sterculia (what species?)

The Rhododendron of the Neilgherries, different
from the Himalayan Rhododendron arboreum in the
colour both of the flowers and of the back of the
leaves.

————

                                                          July 5th.

We visited the Royal Academy for the first time
this year; did not go through nearly the whole, but
stayed long enough to have a good view of some
select pictures. Those which please me particularly
are :—

"Adversity." By *Sant*. A poor looking, thin,
very poorly dressed girl, in black, leaning against a
wall, and offering flowers; very interesting and
touching, the girl's face peculiarly so.

"Esther." By *Long*. Beautiful, and the expression very fine.

"Vashti." By *Long*. Also beautifully painted, but not, to my thinking, equal to the last.

"The Death Warrant." By *Pettie*. The young King Edward the Sixth, very pale and sickly, sitting at the council table with his grave and aged councillors, who are pressing on him the signing of the warrant which lies on the table — he evidently unwilling and shrinking from it; the expressions excellent, and the colouring peculiarly rich and beautiful, quite Venetian.

Portrait of Gladstone. By *Millais*. Very fine.

"The Remnant of an Army." By *Mrs. Butler* (Miss Thompson); the arrival of the one solitary survivor of the Caubul massacre at Jellalabad (in January, 1842). The utter exhaustion, the utterly helpless, hopeless, broken down condition of both man and horse is forcibly expressed.

"Ripening Sunbeams." By *Vicat Cole*. A delightful picture, in his usual style—a scene no doubt in Surrey or Sussex—wide waving fields, golden with the ripe corn, wooded brows and a far-stretching sunny distance.

"The Sea-birds' Resting Place." By *P. Graham*. An interesting view of rocks and sea and sea-fowl, but not as striking as some that I have seen by him.

Millais' portrait of Gladstone has, in even a greater degree than the common portraits, that expression of melancholy, almost of gloom or discontent, which is apparent in all of them. In this

1879. one I might almost call it a look of misery. Norah Aberdare, who knows Gladstone well, thinks the expression is not really that of melancholy, but of habitual intense earnestness of thought. Mr. Eddis, however, says that the lines of Gladstone's face are unmistakably expressive of melancholy.

A visit from my old friend John Carrick Moore, and very pleasant talk. He spent the last winter in Egypt, and his health was very much benefited by the warm, dry climate. He went up the Nile as far as the second cataract. He spoke of the curious Doum Palm, which is peculiar to the Upper Nile, south of a particular degree of latitude ; the singular regularity with which its stem bifurcates; the flesh of its fruit looking like gingerbread, and tasting also like bad stale gingerbread. The Date Palm in the lower parts of Egypt having always a single and solitary stem ; but in the more southerly and hotter parts, growing usually in tufts, several stems of various size growing apparently from one root. The granite quarries of Upper Egypt or Nubia, unfinished columns and statues remaining where they were begun, not yet separated from the solid rock ; some cracked across, probably by earthquake, probable effects of earthquakes visible also in the temples of Thebes.

Carrick Moore has given me several of the fruits of the Doum Palm.

———

July 6th.

Our dinner party on the 5th was very agreeable :

—the Aberdares, Louis Mallets, Mr. and Mrs. 1879.
Oliphant, Mr. and Mrs. Roundell, the Leonard
Lyells, Kinglake, Lady Rayleigh (the Dowager), and
others. I have often before spoken of Norah
Aberdare, and one can hardly praise her too much ;
I hardly know a woman superior to her, either
in agreeable, or estimable, or admirable qualities.
She and her husband are worthy of each other.

Louis Mallet is always worth listening too. Kinglake has (perhaps from ill-health), a curiously quiet
subdued, gentle manner, and I do not perceive
the sarcastic turn for which he was noted at
Cambridge ; perhaps it might appear if I were more
intimate with him. He has a great deal of knowledge and ability.

_____

July 8th.

There seems to be a very general agreement as to
the amiable and estimable qualities of the Prince
Imperial ; he is evidently much regretted by all who
knew anything of him, and his death is looked upon
as a great loss. His will is touching, and shows
excellent feeling, especially the acknowledgment of
the kindness he owed to our Queen and the English
nation. There is still much uncertainty and difference of opinion, as to how far the circumstances
of his death imply error or failure in the officers who
were with him. William Napier and MacMurdo
are decidedly of opinion that the behaviour of
Lieut. Carey and the escort was very blameable.
But the accounts of the affair are so conflicting,
that it is difficult to make up one's mind.

Lord Napier (Napier and Ettrick) came to luncheon with us; he was very agreeable. He said
that he has authentic named portraits of all the
Lord Napiers, his ancestors, and their wives, in
a complete series, from 1600;—that they are all
ugly, and all bad pictures, but of unquestioned
authenticity, and it is very uncommon to find such a
complete series of a noble family. He spoke of the
Scottish painter, Jameson, who was contemporary
with *Vandyck*, and so like him in style, that many of
his portraits have passed for *Vandyck's*. Walter
Scott's amusing story was mentioned, of one of the
Buccleuch family, who, when captured by a feudal
enemy and about to be put to death, ransomed
himself by marrying an ugly daughter of his captor
(known as "Muckle-mouthed Meg;") Lord Napier
said this story had been disproved by documents in
the possession of the Duke of Buccleuch.

————

We had another very agreeable dinner party
yesterday:—Lord and Lady Lilford, Courtenay and
Lady Muriel Boyle, the Clements Markhams, the
Cyril Grahams, my cousin Caroline (or Catty)
Napier, and a few others.

Lord Talbot de Malahide told Fanny, that the
walls of the City of Tarragona in Spain exhibit in
their construction the styles of four different ages
and nations—as it were four different *formations*
(geologically speaking) of building. The lowest and
oldest parts of them are of what is supposed to be

Basque or Iberian work; the next of Roman 1879.
building; the following, Moorish; and the newest,
Spanish, though these also are very old.

In the great siege of Tarragona by Suchet, (Lord
Talbot said) the French cut the aqueduct by which
the city was supplied with water from the hills; but
the Spaniards succeeded in discovering an ancient
well, situated in the centre of the town, which
had been covered up, and over which an amphi-
theatre had been built.

_____

July 12th.

The tormenting skin disease (*eczema*) which I
mentioned June 25th, has gone on without inter-
mission, indeed has grown rather worse than better;
in one respect much worse, for it now torments me
and breaks my rest during much of the night, which
it did not in its earlier stage. The doctor still continues to assure me (and indeed I hear the same in-
directly from other medical authorities) that it is the
effort of nature to throw off morbid matter from the
system, and that if it had not broken out thus, I
should prabably have had a much worse illness.

It seems there is no remedy but patience. The
malady is said to be very prevalent at present.

The continual rains of this extraordinary season
have one favourable effect on the aspect of London;
the grass in the parks is as green as in May; and
the foliage of the trees—the planes especially—is
beautifully fresh and green, not at all soiled or
dingy—the soot having been continually washed off
by the rain.

Dining yesterday with Katharine, we met Sir
Joseph Hooker, also Sir Henry Barkly, who was
formerly Governor of the Cape of Good Hope, is a
zealous botanist, and was a correspondent of Sir
William Hooker. He is a pleasant looking and
pleasant-mannered man, evidently very intelligent
and observant. He mentioned what was quite new
to me, that the Proteas, which disappear on the
eastern frontier of the old Cape colony, and had
been supposed not to range further in that direction
do, in fact re-appear in Natal, where there are table-
topped mountains of similar character to those of
the Cape. Unfortunately, I forgot to ask whether
they were of similar mineral character.

I remarked to Joseph Hooker, that Cape
Proteaceæ have now almost entirely disappeared
from English gardens, though in the early times of
the *Botanical Magazine*, many seem to have been in
cultivation. He agreed with me, and said that they
had disappeared even from Kew, where he is now
doing his utmost to re-introduce them. The
reason (he said) why they can so seldom be kept
alive in modern gardens is, that they are *over-
watered;* the lavish and indiscriminate way in which
watering is practised by modern gardeners is destructive to these and to many other plants of the
Cape and of Australia. The supply of water to
gardens in general being greater and more easy than
in old times; gardeners (and especially garden
lads), get into the habit of being lavish of it: and
much of the care of modern gardeners is bestowed

on ferns and orchids and tropical plants, which 1879.
require much water : still more on parterres and
ribbon-beds and the like. Hooker remarks on the
deterioration of gardening skill in recent times—the
passion for parterres and ribbon-beds, which require
a knowledge of only a (comparatively) few kinds of
plants, and the prevalence of villa-gardening—
leading altogether to the bringing forward of a
great multitude of gardeners of very mediocre
skill. Neglect of study by young gardeners.

Hooker says that the Æsculus (or Pavia) Indica,
of which I gave young plants as well as seeds to
Kew, does not thrive there : the soil is too poor
for it. The soil of Kew is "wretched," he says.
At Nynehead (near Taunton), on the other hand,
the Indian horse-chesnut of which we gave a seed-
ling plant to Mr. Sanford, is growing superbly.

<hr />

July 17th.

Minnie's afternoon party.

Darling little Charlie (aged 4½) in great beauty
and immense glee.

Some interesting talk with Mr. Sanford, and with
Mr. Blunt, the clergyman, formerly of Windsor.

Mr. Sanford sympathizes with my love of trees,
especially exotic ones. He says, he had formerly in
his garden at Nynehead, some myrtles, growing
in the open air, as high as the drawing room we
were in : but they were killed by one peculiarly
severe winter, and now myrtles will not flourish
there in the open air. Neither will apricots ripen

1879. there in the open air. He seems to think the climate is changing for the worse. Perhaps it is only that we are in an unfavourable part of a *cycle* of seasons. The Paulownia will not succeed at Nynehead. He has tried a great variety of Conifers, and finds that the Pinus insignis marks as it were a limit between the many which will bear the climate and the many which will not, as it just enters into the latter class.

Mr. Blunt has lately travelled through Egypt and Palestine. Complains of the want of verdure in Egypt—the universal sand-and-stone colour—the total absence of green in the landscape. (I should feel this very much, for I never could enjoy a landscape without grass or trees. Melancholy impression produced in Palestine by the general aspect of desertion and decay, the prevalence of ruins and deserted villages, the scanty and miserable-looking population of an evidently fertile country.

———

July 25th.

Barton.

We returned to this dear home on the 22nd: my nephew Cecil and Louisa MacMurdo with us; had a good journey, arrived safe, and found all well here. Thanks be to God.

The low lands in the valley of the Stour, both below and above Sudbury, are flooded to a great extent, so that the train seemed, for a long way together, to be running through a sheet of water; trees, sheds, and even cottages, standing insulated.

It is a mercy the railway itself had not been mined and washed away. In East Suffolk, it seems, the railways have in several places been interrupted by the floods.

Here, as elsewhere, the prospects of the harvest are very gloomy, and the hay crop is almost hopelessly ruined. Things indeed are beginning to look better since we arrived ; there has been no rain yesterday or to-day, and now, in the latter part of the day, the weather is serene and fair, so that we may feel some hope. The foliage of the trees is very rich and luxuriant, and the gardens in great beauty of flowers (Roses especially), though with little promise of fruit. I perceive very little damage done in either garden or arboretum by the long winter. Indeed, long as it was, there was no extraordinary severity of frost.

Even the Paulownia, which I fully expected would have been killed by the winter, looks very healthy and flourishing, with abundant and vigorous foliage ; it has not however flowered. The magnolias (grandiflora — trained against walls and protected), look healthy. The Bays and Arbutus perfectly untouched.

Catalpa.—In good foliage, and with good promise of blossom, but remarkably backward.

Æsculus (Pavia) Indica. — Looking well and flowering pretty plentifully, but the thyrses of flowers smaller than I have seen them.

July 26th.

My tormenting malady (from which I have not yet by any means got free) and the incessant bad weather, very much marred my enjoyment of London, and rendered this *season* much less agreeable than any I have experienced for many years. Actually, during these two months (very nearly) I was not once able to visit the Zoological gardens, which always are among my principal attractions in London ; nor to go to Kew. I was able to visit the Royal Academy only once, and the National Gallery once. If it were not for Society, I should say that my London experiences of '79 were nothing but failures. Of society, certainly I had enough ; and not only of *parties*, of strangers and slight acquaintances (for which I care little), but comfortable, pleasant talks with friends whom I esteem and value, as well as with those more intimate connections whom I really love.

The last news from South Africa is satisfactory, it clears the fame of Lord Chelmsford ; casts a new lustre on our army, and best of all gives hope that we may soon hear of peace.

———————

July 30th.

A blessed change of weather since we came home. Of the eight days since our return, five have been entirely without rain, and the last four, 27th, 28th, 29th and to-day—have been serene and beautiful, very warm—real summer weather. Of course the haymaking has been carried on with great zeal, and we have renewed hopes of the harvest. It will 1879. indeed be a mercy if the weather continues favourable.

The roses here are now in great beauty: the season seems a good one for them.

Very pleasant company in our house in the last days: Mr. and Lady Muriel Boyle, Lord Talbot de Malahide and Miss Talbot, Mr. and Mrs. Locker, the Barnardistons, Jos. Rowley, Colonel and Mrs. Ives and others.

Mr. Locker (rather celebrated for his light poetry), and Mrs. Locker are new acquisitions to our list of acquaintance, and very pleasant ones: the lady remarkably attractive and agreeable, lively and well read: and she won my heart at once by her enthusiasm for botany. Mr. Locker is also very agreeable: seems fond both of art and literature, learned in pictures, china, &c.

Lord Talbot remarkably well read, especially in historical literature and anecdotes—also in archæology—a ready talker.

Scott came to give me an account of the mischief done by the storm of the night before last. The most serious is in Mildenhall Fen, where there is great reason to fear that the river will have overflowed and flooded much of the fen (though no report has yet come of this having actually happened). At

1879. West Row, Mr. Gittus's farm house was struck by lightning, but no further damage was done than the knocking down one chimney and melting all the bell wires. Great alarm was caused all through the western part of the parish.

Here, at Barton, all the mischief done was by the wind, as the lightning seems not to have come very near: at least I have not yet heard of anything being struck. An old and remarkably fine walnut tree close to Mr. Baldwin's farm-house, the finest tree of its kind in the parish, was blown down, and in its fall destroyed a shed in which there were some horses, but these fortunately escaped. A large tree also was blown down across the road near the "Bunbury Arms."

Perhaps the storm was something of the nature of a cyclone: at least, here, when it was at its height, the wind is said to have changed with extreme suddenness from the E. to the S. W.

(August 4th. Afternoon).—I am very glad to find that no serious damage appears to have been done in our arboretum by the storm: in fact I do not see that any tree which I value much has suffered at all, except the tulip trees, and those, though many twigs have been broken off, are not perceptibly disfigured.

————

August 5th.

The Bury paper full of particulars of the storm of the 3rd, and the extensive damage done by it in numerous places in this neighbourhood. It seems to have been more violent and destructive almost everywhere around us than just here, as if we had escaped by being in the centre of the orbit in which it revolved.

Dear Joanna Horner, with her and our niece Dora, arrived. Joanna has not been here since '73. She is looking well and seems in good spirits.

August 6th.

Scott, who went to Mildenhall yesterday to inspect the state of the Fen, reports that the river has not yet actually risen over the bank, but that the danger is not yet over, because there is yet a good deal of water which may come down from the higher ground : and the soil of the fens is altogether in a very *sloppy* and insecure state. He saw, on the Icklingham plain, the water actually standing in pools on the surface of the sand.

August 9th.

Dear Cissy and Emily arrived, both looking very well.

August 11th.

Our dear sister Joanna and Dora Pertz set out on their journey to Scotland.

August 12th.

Beautiful weather the last four or five days. This change of weather has been a great blessing. The hay harvest has gone on most prosperously in

1879. these latter days; all the grass in the park has been cut, and nearly all of it will be on the waggons by this evening.   Scott tells me that, including that now on the waggons, we have enough hay to make a first-rate stack, and in as fine condition as possible.   There is another large stack at the farm, but part of this, having been cut too early, is not of such good quality.

August 13th.

The unfortunate people of Lakenheath have again had their Fen overflowed, the embankment of their river (the Little Ouse) having given way.   The same misfortune happened to them only a few years ago.   Mildenhall Fen has been spared, but has had a narrow escape.

August 17th.

Fanny has had a description of the wedding of Frank Lyell to Miss Guise in a letter from Katharine, who was present at it.   I wish her and the young couple much happiness.

The hay in the park here is not *yet* all carried, though only a small portion (chiefly on the north side of the house), remains out.   It is remarkably late, but Scott considers it a fine crop and in excellent condition.

August 25th.

Dear Cissy and Emmy left us on Saturday,

having been with us a fortnight. I do not know 1879.
a better woman than Cissy, a truer and more loving
heart, or one more devoted to duty.

The guests of the week all went away, except
Mrs. Ellice and her daughter Helen. The party
had been (besides those just now mentioned), Lady
Louisa Legge, Susan Hervey, Mr. and Mrs. Cyril
Graham, Mr. and Mrs. Calliphronas, John Hervey.
Three of these ladies are Herveys ; three sisters, all
very agreeable, and very different. All the ladies of
the party I should say are decidedly agreeable.

Mr. Cyril Graham is a remarkable and interesting
man; a prodigious traveller and a prodigious lin-
guist ; acquainted with I know not how many
countries, east and west, north and south ; has
experienced the extremes of heat in Central Africa
and of cold in North America ; has observed an
immense deal, and talks readily, willingly, and well
of what he has seen.

August 26th.

Arthur went up to London with his tutor Mr.
Trevor, to go in for his "Preliminary" examination
for the army. I heartily wish him success.

August 27th.

Mrs. Ellice and her daughter Helen left us. They
are both very pleasant. We are now quite alone (a
wonder !) Raining furiously the whole day. The
prospects of the harvest are most dismal, and all the
low parts of the country are flooded. I remember
nothing like it since 1860.

1879.   Sir Rowland Hill, one of the most memorable
men of our time, died yesterday. Few men have
done more important service to his country, perhaps
to mankind, than the author of the Penny Postage.
And this I say in spite of the abominable nuisances
which that postage has brought on us in the shape
of circulars, advertisements, reports and begging
letters.

September 1st.

Sir Rowland Hill is to be buried in Westminster
Abbey : few men more deserving of it.

September 2nd.

Yesterday was a splendid day ; brilliant sunshine,
with a fine, cool, fresh, exhilarating breeze. Wheat
harvest begun on my home farm, and generally
in the parish. Farming prospects looking rather
more hopeful.

Lord Hanmer, in a letter to Fanny, mentions
that an Araucaria Brasiliana, standing near his
house at Bettisfield, twelve feet high, has borne the
whole of the last winter without the least artificial
protection, and is now flourishing. It is indeed, he
says, in a very warm and sheltered spot, protected by part of the house, but still it is very
unusual for this species to bear so much cold.

September 4th.

Beautiful weather ; the harvest going on briskly.

Scott, however, tells me that an unusually large 1879.
proportion of the wheat is attacked by mildew : and
moreover, that the more luxuriant, the more highly
cultivated is the wheat plant, the more it suffers
from this fungus.

September 5th.

We spent a very pleasant afternoon yesterday at
Livermere : drank tea with the Hortons, and
strolled through the gardens; the MacMurdos
(including Mimi), with us.

There are very fine trees at Livermere, particularly two or three magnificent cedars, and a
very grand Elm (Ulmus glabra ?), which, as Admiral
Horton tells me, measures 16 feet 9 inches in
circumference at 3 feet from the ground. All these
are on the garden front of the house—the front
which looks to the water ; the other side, to which
we drive up, is very poor-looking.

September 6th.

The MacMurdos left us; I am very sorry to part
with them. I always enjoy their society much—
Montague's especially. Emily (Mimi) is lovely and
charming.

Received a letter from Dudley Hervey, from
Singapore, enclosing a specimen of a very curious
and interesting fern, which he had found on the
highest mountain in the territory of Johore.* I

* Johore is the extremity of the Malay Peninsular, nearest to, and opposite
to, the island of Singapore.

1879. have since ascertained it to be the *Acrostichum bicuspe* of Spec. Fil. *Gymnopteris Vespertilio* of Hook. Lond. *Journal of Botany*, v. 5. t. 7.

———

September 8th.

One field of wheat on my Home Farm has now been cleared, the corn carried and placed under shelter : but it must be kept some time before it is thrashed, being still in a very damp state. The same is the case with the wheat in the other farms in the parish : it takes long to dry. The wheat on my farm is, it appears, of excellent quality.

Terrible news from Caubul.

There seems no doubt that the brave Cavagnari and all his staff and escort have been massacred; there is hardly a hope of their escape.

So now we are in for another Afghan war ; probably a much more severe and bitter one, lasting longer, and involving a much greater expenditure of precious lives as well as of money. The resemblance to the old disaster of Burnes and MacNaughten is striking. It is true that our acquisition of the mountain barrier, with its passes, in the late war, has given us a better base of operations, and enabled us to start on our new invasion within a much less distance of Caubul, and with fewer difficulties to overcome. But still I cannot help thinking that we should be in a better position if we had left Shere Ali alone, and not interfered with so wild, lawless, and desperate people as the Afghans.

September 11th.

Archdeacon Chapman reminded me yesterday, that the great objection made by the late Ameer, Shere Ali, to the treaty which our Government wanted to force upon him—the objection, in fact, upon which the whole broke off—was, that he could not sanction the establishment of an English Embassy at Caubul. He said, he could not guarantee their safety—could not undertake to guard them from the mob of that city. He knew his own people, it seems, better than we did.

The dinner party at the Chapmans was very pleasant:—the Bishop, Agnes Wilson, Augustus Phipps and his daughter, Mr. and Lady Florence King, and one or two others. The Bishop is remarkably agreeable.

Mrs. Chapman showed me some interesting MSS. of Southey, particularly a volume of miscellaneous fragments written by him while at school. She is a niece of Southey's great friend, Grosvenor Bedford.

————

September 19th.

We returned yesterday from a three days visit to Lord and Lady Tollemache at Helmingham: very pleasant. It was the first time we have been there.

Helmingham is a fine and interesting old house, very complete and well preserved in its archaic character: enclosed (almost completely) by a broad and deep moat lined with brickwork, and accessible by a drawbridge, which is raised every night. The house (built, as I understood, in the time of Henry Eighth), is of very venerable and handsome-looking

1879. red brick ; built in the form of a square, enclosing a large square paved court. We enter this court immediately from the drawbridge, and from the court we pass into a fine old hall, large and lofty, adorned with stags' heads, armour, and old portraits. All the rooms (that I saw at least), are quite consistent in their style of fitting and furniture ; ornamented ceilings, walls pannelled with very dark oak ; chimney pieces in the same style, very high, very dark, and very magnificent ; mullioned windows and everything in character. The library is not magnificent, but a handsome antique-looking room, with plenty of old books and many interesting specimens of typography and manuscripts, which Lord Tollemache very kindly shewed us on the 16th. A fine copy of Caxton's "Game of Chess," the very first book printed in England. (I have seen it also in the British Museum). Several MS. books of devotion, exquisite specimens of writing and illuminating ; some of them perfectly wonderful for the minuteness, regularity and delicacy of the hand writing, examples of almost incredible industry and patience. The beauty and perfect preservation of the paper, and the fine quality of the ink, are very remarkable. Queen Elizabeth's lute and spinnett, given by herself to Sir Lionel Tollemache, who had entertained her here in 1561, are preserved in the house. The lute is prettily ornamented with delicate inlaid work. The spinnett appears to have been a sort of small pianoforte, very primitive.

The garden, lying outside of the moat, but enclosed by a moat of its own, is pretty, in the old style, with broad walks of green turf, and yews 1879. carefully clipped into conical shapes. ; all very neat, trim, and old-fashioned; plenty of handsome flowers, but nothing botanically noticeable.

The church on the edge of the park, within sight of the house : — very like, externally, to that of Barton ; the tower particularly, but the nave longer. Within, very numerous monuments of the Tollemache family, several old and quaint. The one which most interested me was that of the General Thomas Tollemache (Talmash, Burnet spells the name), who was killed in the attempt on Brest in 1694 ; this tomb has a very long inscription in English, in which it is stated that in the fatal affair of Brest, he was " misled by pretended friends."* (I think these are the words, certainly the sense)—thus glancing at Marlborough without naming him.

A huge monument in the style of James First's time, containing in four niches kneeling figures " of the four first Tollemaches who settled at Helmingham."—*Murray*.

Another remarkable monument of a Sir Lionel Tollemache ; a graceful and well executed figure (coloured), reclining on one side, with his head resting on his hand ; the dress and beard appear to be those of Charles the First's time.

Park very extensive, not picturesque (as far as I saw) except by reason of its abundance of huge and venerable oaks.

---

* Lord Tollemache told us that these facts were all preserved in the records of Paris.—F. J. B.

1879.   Red deer—a fine herd in the park, of which Lord
Tollemache got for me a very good view—fine
stately stags with grand antlers.   There was a herd
of fallow deer grazing near them at the same time,
a very good opportunity of seeing the contrast, but
he said, they never intermix.

The house has a high roof—no conspicuous
parapet.

Pictures very numerous, chiefly portraits more or
less connected with the family, and of more
historical than artistic interest.   A portrait of Mary
Tudor, " Bloody Mary " as a child ; not at all an
unpleasant face.

A very curious little picture, containing full
length portraits of Henry VIII. and Queen
Elizabeth, *both* represented as *full-grown*, and in
robes of state, as King and Queen ; the whole
elaborately painted like a miniature.

On the 17th, Mr. Cardew, the clergyman (a very
intelligent and pleasant old gentleman), showed me
a very curious collection of Roman-British antiquities, principally broken pottery, which he had
discovered near the Church and in the grounds of
the Rectory.   I have not time just now to describe
them.

Lord Tollemache has a very large family ; there
were, I think, eight sons, and the one daughter at
home during our visit : and there are besides two
sons of the first marriage.

The guests together with us were :—Mr. and
Mrs. Roundell (who left us at Barton in the
morning, and met us again at Helmingham), Lord

Hatherleigh (for the first day), Mrs. Leicester and Mr. and Mrs. Lowe. I had long been rather curious to meet Mr. Lowe (the celebrated *Bob Lowe*), and was uncertain how I should like him : but he made a decidedly pleasant impression both on me and Fanny. I suppose the company was agreeable to him, for I did not observe anything bitter or cynical in his conversation. He spoke indeed with severity of the aggressive foreign policy of the present Government, especially in Afghaunistan : but I thought his observations very just.

LETTER.

Mildenhall,
September 24th, 1879.

My dear Edward,

You were curious to know what impression Mr. Lowe would make on Fanny and me, and from what you yourself have said to me, I daresay you will not be surprised to hear that the impression was decidedly a pleasant one.

I suppose his surroundings were to his taste, for I observed nothing harsh, bitter or sarcastic in his manner or conversation ; he talked on a variety of subjects easily and pleasantly, and, as I thought, wisely ; his remarks on the Afghaun business, in particular, I thought very judicious. Fanny, who had more of his conversation than I had, took rather a fancy to him. The only other guests were the Roundells, Lord Hatherley (only for the first day), and Mrs. L. Lister, (Leicester ?), whom we

1879. had met last year at Merton. Altogether we had a
pleasant visit at Helmingham. Do you know the
place? It is a very fine old house, of Henry
Seventh or Henry Eighth's time: very complete
and uniform in its style, completely surrounded by
a broad moat, over which is a draw-bridge, which is
raised every night: a large square courtyard in the
centre. A library full of quaint and curious old
books, with several valuable MSS., and early
printed books in fine condition; a herd of red deer
in the park (we had a very good view of them), and
many magnificent old oaks.

We came hither on Tuesday, the 23rd, a day of
excessive rain, and must stay here probably till the
30th. Joanna Horner and Dora Pertz are with us,
as well as Arthur and his tutor. By-the-bye you
will not yet have heard that Mr. Arthur passed his
preliminary examination for Sandhurst successfully,
a great satisfaction to us: we heard the news while
we were at Helmingham; he is to go up for his
qnalifying examination in December.

The new plantations here are thriving: the wet
season has agreed better with them than with
anything else; nothing can be more deplorable than
the accounts I receive of the state of the crops and
the farms here. The farmers are most doleful,
and no wonder; 1 believe a good many of the
smaller ones will be unable to pay any rent at all
this autumn, at least for their fen farms, and my
income this year will certainly be seriously reduced.
But only one, as yet, has made any complaint of
his rent.

I hope you have enjoyed your tour in France, and 1879. that it has done you good. I look forward with great hope of instruction to your "Ancient Geography."

Fanny sends much love.

Ever your loving brother,

CHARLES J. F. BUNBURY.

September 28th.

JOURNAL.

October 1st.

We returned from Mildenhall yesterday, and this morning I see in the local paper the death of our dear old friend Mr. Mills, the rector of Stutton. It is not to be regretted on his own account, for he was not only very old (if I understand rightly, he was 86 when we were at Stutton last year), but had for some time been very infirm, and had latterly become quite blind, after being nearly so for some years. For himself, I should think death must have been a release : but I grieve for his excellent wife, who was quite devoted to him. He was a very good man, most amiable, kind-hearted and friendly, and a very pleasant man ; not perhaps a man of deep or extensive learning, but with a great deal of miscellaneous knowledge, and especially full of anecdote, having known a great variety of people, and bringing out his knowledge of them in an easy, pleasant, chatty way : never reserved and never overwhelming in his talk. His knowledge of genealogy, family history, county and local history,

1879. was indeed remarkable; scarcely anybody connected with Suffolk could be mentioned, but what Mr. Mills knew something about his family and personal history.

Our acquaintance with Mr. Mills began in 1852, and in '58 we first stayed as guests at Stutton, and from that time we visited them almost every year, and our acquaintance grew into a fast friendship. It was a particular pleasure to see Mr. Mills in his beautiful home at Stutton; he had such hearty enjoyment of it, and did the honours of it so pleasantly and with such evident satisfaction. Mr. Mills appeared to me to have a delicate and refined taste in all the decorative arts, including those of gardening and laying out ground, in which Fanny and I often consulted him. His wife has, I should think, a stronger intellect than his (though less of acquired learning), but they always appeared thoroughly congenial, and "if she ruled him," she "never showed she ruled."

She is indeed very very much to be pitied.

To us Stutton can never again be the delightful place it has been.

October 5th.

We went to Mildenhall, the 23rd September, and returned hither on the 30th. We were very busy all that week visiting tenants and receiving their visits, visiting schools and churches, and looking at some of the new plantations. I saw six or seven of the principal tenants, and two or three of the small farmers; all of them, as might be expected, were

very doleful and desponding, especially as to their 1879.
fen farms, but only one (a new tenant) complained of
his rent ; all the rest were very civil, though several
of them said plainly, that unless next year were a
much more favourable season, it would be impossible
to go on farming.

In other respects our visit to Mildenhall was more
cheering. We found house and garden in excellent
order and condition ; our old servants, the Elmers
and Agnes Fincham, doing their work admirably,
and seemingly glad to see us ; our boys' school most
satisfactorily conducted by that excellent man, Mr.
Saxton ; the Vicar, Mr. Livingstone (the second
Vicar whom I had presented to the living) very
popular, very cheerful, earning golden opinions
from all sorts of men, and comfortably established
in the new Vicarage house which has been provided
at Mildenhall.

We visited the new Church (St. John's) which has
been built at Aspal, between Beck and Holywell
Row ; we were very much pleased with its appearance and fittings up, and we hear that it is very well
attended, and is much appreciated by the neighbourhood.

We were present at the ceremony of the foundation of this church. Fanny *laid the first stone*, as the
phrase is.

The subscription for building this Church was set
on foot, by Mr. Robeson, and it was mainly by his
energy and persuasive powers that it was successful ;
he also, himself, subscribed largely, as of course we
also did. His energy and power of getting others·

1879. to promote his objects, are indeed extraordinary. We visited also the little *iron* church at Kenny Hill, built by Mr. Robeson, entirely at his own expense; this also is in very neat order, and is said to be well attended, though there are very few habitations within sight of it.

I think it is an extraordinary blessing of Providence, that I should have been allowed to appoint *two* such men in succession to the living of Mildenhall, as Mr. Robeson and Mr. Livingstone. When the first of these went away to Tewkesbury, I almost despaired of finding a fit successor to him, but Mr. Livingstone has in these two years proved himself quite worthy to fill the place, and not only useful, but agreeable to the parishioners of all varieties of opinion as well as to us.

Mr. Robeson was recommended to us by Theodore Walrond; and Mr. Livingstone by Mr. Robeson.

———

October 10th.

Lord Charles Hervey came to us on the 3rd, and went away yesterday; his visit was very pleasant to us. He and I harmonize completely in our love of natural history, especially of botany, and I had many long and pleasant talks with him on these subjects; for without being profoundly scientific, he is devoted to natural history pursuits, is very observant, and has very considerable knowledge. His hearty love and enjoyment of nature is delightful. He has been a considerable traveller too, having been sent by the doctors to one distant country after another ; so that he has seen the West Indies, 1879. Rio de Janeiro, California, and a good deal of North America, the Sandwich Islands, Cape of Good Hope, Australia, and perhaps other countries; and wherever he went, he has observed and profited well by what he saw. The last time he was here, if I am not mistaken, he was lately returned from California and the Sandwich Islands, and told us many curious things concerning those countries, which I believe I noted down at the time.

He does not seem to have since visited any distant countries, but he has passed several winters, for his health, on the Genoese Riviera. There he has kept up his love of nature, and his habits of observation, and I much enjoyed his talks on the botany of that beautiful region.

Lord Charles Hervey was very handsome when a young man ; when I first remember him, he had a striking likeness to his brother, the Bishop ; but habitual ill-health has given him a lean, worn look, and made him appear older than he is. He is, I should think, a man of a very sensitive disposition, of a delicate, refined mind.

October 13th.

Dear Joanna and her niece Dora went away. I was very sorry to part with them, especially with Joanna, whose agreeableness as well as her other excellent qualities I have so long known. She had not been in England for 6 years until this time. She is looking very well, young for her age, vigorous and active in mind and spirit.

October 16th.

My Barton Rent Audit; rather a melancholy
ceremony in such a season as this. My tenants,
however, behaved very well, were very civil to me,
and (with one exception), did not complain of their
rents, though they were very doleful (as well they
might be) concerning the weather and the deplorable
failure of the harvest. Two of them were unable to
pay the whole of the usual instalment " on account "
of the rent, but have promised to make it good
by-and-bye. Altogether, my receipts at this Audit
are £210 less than at the same time last year : but I
much fear that next year the deficiency will be
greater. The practice here has been for each tenant
to pay a full half of the year's rent at the Audit
immediately after harvest ; all the deductions and
allowances (taxes, repairs, &c.) being made from the
payment in February.

October 18th.

The good news of dear Kate Hoare's safe
delivery ; she has now three boys.

October 31st.

There has been more fine weather—if I am not
mistaken—a greater number of fine days in this
month than in any other of this year; certainly
there has been no other in which one has so often
been disposed to remark that the weather was
fine.

According to my gardener's register, there have

been twelve days (of 24 hours) in which no rain at 1879. all has fallen: and seven others in which the rainfall has been less than five-hundredths of an inch. On only three days has the rainfall exceeded ·10 of an inch; being ·30 on the 1st of the month; ·25 on the 25th; and ·13 on the 20th.

Much agreeable company:—in the first place, from the 17th, dear Minnie and Cissy and Emily, who are with us now. Next, on the 23rd and 24th, the Bishop of Ely,* a very agreeable man, to meet whom, we invited some of the clergy, in particular, Archdeacon Chapman and Mr. Livingstone. Then the Goodlakes, from the 14th to the 28th:—old friends, whom I have often mentioned. Mr. and Mrs. Wilmot Horton from 25th to 28th; Mr. W. H. a particularly pleasant man. Lastly, our Ball party, of which the greater number are still here:—Lady Winchelsea and her daughter Lady Evelyn, Albert and Sarah Seymour, Lady Mary Egerton and her daughter Georgina, Leopold and Lady Mary Powys, Lina and Sarah and Harry Bruce, Lord John Hervey, Rosamond and Arthur Lyell, Dora Pertz, Josiah Rowley, and my nephew Harry. A good houseful—many of them eminently agreeable.

The autumnal colouring of the foliage has been very fine this month, which I should not have expected after so cold and wet a summer. The colouring of the Japanese Creeper (Vitis or Ampelopsis Veitchii) has been glorious: and the beauty of an American Maple, with the most

* Bishop Woodford.

1879. brilliant scarlet glowing out here and there amidst its deep green, is wonderful

———

We had the sad news, in a letter from Susan Horner at Florence, of the death of Martha Somerville. We had indeed heard, some days ago, that her case was quite hopeless. I am afraid she suffered much in her last illness. So, there, another old friend is gone ; not that she ever was a *very* intimate friend of mine, but she was of Fanny, and I was acquainted with her for between 50 and 60 years,—from the time when I was a mere boy. I received so much kindness from her father and mother, I was so often in their house at Chelsea and again in Rome, and had so much pleasure in Mrs. Somerville's conversation, that I cannot but feel sad and serious when I consider that that family has entirely passed away.

Martha survived her father and mother, her sister and her half-brother, and has left no relations but distant ones. It is a comfort to hear that she was tenderly nursed in her last illness by good servants and kind friends.

———

Dear Mimi MacMurdo left us, by the express wish of her parents, to join them at Alassio. I am very sorry to part with her ; she is not only lovely, but a most agreeable, cheerful, indeed charming companion.

Weather very sharp these last three days : 1879. thermometer down to 26 deg. on the 13th, to 25 deg the 14th, and to 23 deg. this morning. Yesterday was a beautiful day, and to-day still more so : splendid sunshine, almost cloudless sky, and no wind, so that one hardly felt the cold. But these sharp frosts have cut down all, or nearly all the half-hardy plants, such as Cannas, Ricinus, &c., several of which had continued in beauty till a few days ago ; and the frost, together with a high wind just before, have stripped most of the trees, and browned nearly all that remains of foliage (evergreens of course excepted). All the autumal beauty of the gardens and shrubberies is at an end, and the winter has fairly set in.

The re-arrangement of our old fern-house has been completed just in time ; the rockwork (of slags—*clinkers* as they are called), is arranged, and the ferns, planted in the hollows, look very pretty.

November 24th.

Our very dear friends, the Bishop of Bath and Wells, and his wife, with their youngest daughter (Caroline, or Truey, as she is always called in the family) came to us last Friday, the 21st and left us to day. It was a cruelly short visit (they have such pressing demands on their time), but yet a very great pleasure to us. They have never been here since dear Sarah's death (almost exactly two years), indeed, we have not, since then, once seen Lady Arthur till now. Arthur Hervey does not look well,

1879. his appearance shows rather more of the wear and
tear of his anxious and troublesome position. I love
them both dearly.

It is about two months over fifty years, since
I first became acquainted with Arthur Hervey;
when he was 21, and I about 20 and a half. We
made the acquaintance of his wife soon after we
married in '44 ; but (owing to particular circumstances affecting us) it was not till several years after
that we came to know her intimately. Ever since
then we have valued her very highly.

We have had other very pleasant guests in the
house at the same time. My dear cousin Caroline
(Catty) Napier, Lady Muriel Boyle, Mr. and
Mrs. Montgomerie, Admiral Spencer, Mr.
Egerton Hubbard, Miss Sullivan, and Edward.
Nearly all these are old friends, whose merits I have
repeatedly mentioned.

Mr. Hubbard has travelled much, and observed
well, and without being a scientific botanist, has a
great love of plants, and especially of trees, so that
he and I fraternize thoroughly on this subject. Miss
Sullivan, also, a friend of the MacMurdos at
Fulham, a very pleasant person, devoted to botany
and gardening.

November 26th.

Very wintry weather ever since the 20th. On
that day the weather changed, and became bitterly
cold (having been very fine and mild in the first
days of the week). The 21st there was a heavy fall

of snow, which lasted to the middle of the day 1879.
and lay partially till night, indeed it has not since
entirely departed.

December 2nd.

We are now really alone ("Darby and Joan"),
which is very unusual with us.   Our last guests left
us on Saturday, the 29th November ; dear Catty
for Oxford ; Edward for London; Arthur and his
tutor Mr. Trevor also for London to pass (as we
hope) his qualifying examination for Sandhurst.

The weather is terribly wintry ; thermometer
down to 10 deg. Faht. last night, and the ground is
covered with snow.   A severe winter will be a
serious aggravation of the distress which threatens
this part of England.

December 6th.

Arthur returned from London in very good health
and spirits, having gone through his examination ;
it will be a good while, I am afraid, before they will
let us know whether he has *passed*.

Received Edward's book, " The History of
Ancient Geography," two goodly volumes, which
will take some time to read.   He has been long
engaged on it, and I have no doubt it is very well
and carefully done.

December 8th.

I see in *The Times* the announcement of the death

1879. of Sir William Boxall, at the age of 80. We have known him a long time, and from time to time have seen much of him; especially in the spring of '53, when he and we were staying at Ventnor, and used to meet almost every day; and again in '56, when I sat to him for my picture. He was always very friendly and cordial with us. He was, I believe, a very good man, very kind and benevolent; a very agreeable man also, very well informed, and with his memory fully stored with anecdote. He was always (since I have known him) a great invalid, and not a little of a hypochondriac; and (when we had come to know him well) it used to be rather amusing to hear his dismal, desponding accounts of himself, and to observe how he rose out of that state of gloom as he went on talking. He would begin when we first came in to him, with talking of his complaints and sufferings, as if he had nothing to expect but speedy death; then, by degrees, and not very slowly, as he went on talking, he grew more and more cheerful, till the gloom was completely dispersed, and his talk became genial, various, animated and delightful. His health, I believe was really bad, and he constantly talked as if he expected his death to be very near; yet he has lived to eighty.

Two of the best pictures by Boxall, which I have seen were the portraits of Gibson, the sculptor, and of our friend Lady Cullum.

———

December 15th.

Our dear Arthur MacMurdo set out for London,

where he is to join my nephew Cecil, and make a 1879.
rapid run with him to Florence and Rome. He
starts in high health and spirits; may God protect
him and bring him back safe and well to us.

———

December 15th.

A welcome change of weather, and what makes it
more welcome is, that the thaw has been very
gradual, so that there is less danger of floods. Ever
since the end of last month the frost has been
steady, and the ground covered with snow, which is
only now disappearing.

The news from Afghanistan in these last few
days, has been very distressing. The severe
fighting near Caubul, against a very great force of
the enemy—with the loss of six officers killed and
ten wounded — shows that we have to deal with a
formidable uprising against our authority, not a
mere émeute : and I very much fear that a great
many valuable lives will yet be sacrificed in this
miserable enterprise. Being so far engaged in it, I
suppose we must go on till we have gained some
striking success : for an apparent failure might be
dangerous to our reputation in India, and therefore
to our position there.

We had, last week, a pleasant three days' visit
from Mr. Eddis, the painter; he came on the 10th
and left us on the 13th. He is a remarkably
pleasant man, in spite of the deafness which has
grown upon him : very cheerful and genial, and
well informed. I am not sure whether I have

1879. mentioned before, that, while we were in London in the summer, I, by my wife's particular wish, sat to him for my picture: and he painted a portrait which was very satisfactory to me and to those friends who have seen it. When he was here the other day, he added some finishing touches to this portrait, and also executed for Fanny a very capital sketch of Arthur.

December 21st.

We have had very comfortable accounts from Arthur and Cecil on their journey: on the 18th from Paris, and this morning from Marseilles, where they had arrived safe and sound ("as merry as clowns," Arthur says), and without having met with any impediment.

LETTER.

Barton Hall, Bury St. Edmund's,
December 22nd, '79.

My dear Edward,

I heartily wish you a happy Christmas and New Year and many of them. I suppose you will soon be leaving town to enjoy these holidays with some of your friends in the country, and I hope you have not caught and will not catch any cold. My chief purpose in writing (besides good wishes), is to tell you that I have begun upon your big book, have read as far as the end of the section on the Greek colonies, and like it very much. It is certainly not *light* reading, but *that* was not to be expected : it is so full of matter that I cannot read it fast, for it requires close attention, but I find the matter excellently well arranged (which is a most important point), and the style clear and pleasant. As far as I have read, I find no difficulty except what is inherent in the nature of the subject and my imperfect knowledge of it. I am interested by your discussion on the Phœnician voyages, but it is new to me that *amber* is rare in Sicily; I remember our father and mother brought some from thence, and I fancied it was plentiful. Your section on the wanderings of Ulysses entertained me much, and I think you have pretty clearly made out that Homer did not pretend to give a true geographical account of a real voyage. But I expect you will have Mr. Gladstone *down* on you if he has any time to spare from abusing the Ministry. The section on the Greek colonies gives a wonderful idea of the spirit and energy of the Greeks, even in those early days.

We are now *quite* alone, an uncommon thing with us, and shall be quite alone at Christmas, which we have not been for the last ten years.

Arthur (I suppose you know), has gone with our nephew Cecil to see Rome—as much as he can see of it in the time, in his holidays, and we have very pleasant and comfortable accounts from both of them from Cannes.

We are both of us well, I am thankful to say, in spite of this terrible season.

(December 23rd).—Many thanks for your letter,

1879. which I have received this afternoon. Our letters would have crossed if I had not been interrupted yesterday by George Betts, who came to talk to me about Mildenhall business.   I am very glad you are well, and hope you will much enjoy your visit to Kingston.   Many thanks also for your good wishes. Very glad your book is approved (as I did not doubt it would be), by those whose approbation you value.

Poor Lord Arran's is indeed a sad case, and the worst of it is that I fear he suffers very much.   At the age of 78, I should suppose his recovery from such an illness very improbable.

With much love from Fanny.

Believe me ever,

Your very affectionate brother,

CHARLES J. F. BUNBURY.

JOURNAL.

December 24th.

We have had a succession of most comfortable and pleasant accounts from our travellers, from Paris, Marseilles and now from Cannes, where they have made a little halt (Cecil's mother being there), and seem to be enjoying it thoroughly, after the severe journey from Paris.   Arthur writes in the highest spirits.   They described the cold at Paris as terrible, but at Cannes they have brilliant sunshine, and have been sitting out-of-doors, under the orange trees, eating the fruit.

A delightful letter yesterday from dear Rose 1879. Kingsley, from Leamington, where she and her mother are settled for the present. Her sister, Mary Harrison lives near (at Wormleighton, of which parish Mr. Harrison is the incumbent), near enough for her to go into Leamington and see Rose once a week.

The weather seems to be still more severe than with us. She says, "Grenville," her brother, "wrote " from Edgehill, saying they had had 29 degrees of frost "the night before."

Mrs. Kingsley had had a threatening of severe illness some time before, so as to alarm them, but happily she had rallied and Rose wrote cheerfully, and in good spirits about her.

The accounts from Caubul still not comfortable, though not absolutely disastrous. It is evident that the country in general is up in arms against us, our troops enormously outnumbered, the several detachments and divisions of our force surrounded and seperated from one another by the enemy, communications uncertain ; and that it will be difficult for the reinforcements to join the troops already there. Still, our soldiers are excellent, and under very good officers ; and I hope and believe that we shall ultimately be successful, but at a lamentable cost of bloodshed and misery, a lamentable loss of valuable lives to ourselves, and cruel sufferings to the women and children of the Afghans.

LETTER.

My dear Susan,

1879.   This letter is intended to reach you in time for your birthday, and to wish you many happy returns of it, as well as happy new years, but I am not sure whether I shall finish it in time; but whether I do or not, I equally send my best wishes with all my heart to you and all of your home circle. We are quite alone this Christmas—regular Darby and Joan fashion—which has not happened for very many years, but our servants had a good cheerful party yesterday. It seems rather against the grain to talk of a *merry* Christmas, with such a bitter season and so much distress in town and country, and this miserable, unjust war in Afghanistan; so many hearts tortured by anxiety for their dearest friends. Cissy is very brave, but she cannot help feeling deep anxiety, with her son George hurrying to join his regiment (the 9th) which is in the thick of the fray, and her son Willie also expecting and longing to be ordered to the front. Fanny and I have reason to be thankful that we are well as yet, in this terrible winter, and that the bad times have not as yet reduced either our comforts or our means of helping others. I try to believe in Kingsley's doctrine of the healthiness of a severe winter like this (what is called seasonable weather), but I cannot bring myself to enjoy it. In obedience to

the precepts of my two doctors (Dr. Andrew Clark 1879.
and Dr. Fanny, I take a walk every day; I grumble,
but I walk. The glowing accounts which Arthur
and Cecil write of the beautiful weather and the
sunshine on the Riviera, make me almost envious;
but I console myself in our greenhouses, which have
even now a great deal of beauty in them, as well as
a genial temperature. We have plenty to do, plenty
of books, and a good house, and are as happy as
anxiety for absent friends will allow.

This year has been hitherto a rather specially
happy one for us, inasmuch as we have lost fewer
intimate friends by death than in several previous
years; indeed I think none, except dear old
Mr. Mills. Those losses are what (as Gibbon says)
cast a browner shade over the evening of life.

Before you get this letter, you will, I suppose,
have seen our travellers, Cecil and Arthur. We
have had the comfort of receiving regular reports of
every step of their progress, and very joyous and
happy reports, which make Fanny very happy.
I will leave her to tell you what is to be told about
us and our belongings, neighbours, servants, animals,
&c. &c., and proceed to a little talk about books :—
I have begun to read Edward's big book (a big book
indeed it is), have read about 150 pages, and like it
very much. Of course it is not light reading, nor
suited to the taste of "the *general* reader," but it is
written in a good clear style, and the matter is very
well arranged. It is full of curious matter, and I
have not the least doubt that it is scrupulously
accurate. He is pleased by the remarks he has

1879. heard made on it by those who are qualified to judge.

I have been much interested by the third volume of Green's "History of the English People;" which I am just finishing : the style indeed is strained and artificial, rather fatiguing to read, but it is very interesting, and I like the man's sentiments and opinions. There is not so much in it which is new to me as in the previous volumes ; for I have read a good deal about the reigns of the two Charles's ; but it is a period of our history which I should not grow tired of reading. In the evenings, now we are alone, I have begun reading to Fanny, Kaye's " History of the former Afghan War."

Pray give my best love to dear Leonora and Joanna, and tell the latter I often think of the pleasant days while she was here ; and once more, I will heartily wish a happy new year to you and all your home circle. May 1880 be fruitful of blessings to you all.

Ever your loving brother,
CHARLES J. F. BUNBURY.

JOURNAL.

December 27th.

Luculia gratissima has been in blossom in the conservatory here for a month. It is very beautiful and very fragrant. The plate in the *Botanical Magazine* by no means does justice to the delicate beauty of the flowers.

The terrible news of the breaking down of the bridge over the Tay, near Dundee, and the consequent destruction of a whole railway train with all its passengers. It is an awful catastrophe.

The same papers give us the news of Sir Frederick Roberts's great victory near Caubul, which will, I should hope, be decisive and lead to a speedy end of the war.

LETTER.

Barton, Bury St. Edmund's.
December 29th, 1879.

My dear Katharine,

We are very near the end of the year, and with all my heart I wish a happy new year and many of them to you, and all your home circle. I am very sorry that you could not come to us at this time, but very glad that you will have such a pleasant, cheerful, family party at home instead.

Many thanks for your kind messages to me, and for the pretty pen-wiper you sent me.

We have had delightful accounts of Arthur to day from himself and his family, with whom he seems to have passed his Christmas very happily, all seeming mutually pleased. He and Cecil are I suppose now at Florence.

I hope that the successes in Afghanistan, which are related in to-day's paper may lead to a speedy close of the war, and that the fighting may be over before my nephew George reaches the frontier. It

1879. is a wretched business at best, but we can now only get out of it by victory.

I wish you could have seen our *Luculia* while it was in beauty, but perhaps you know the shrub; it is a beautiful thing, closely allied to the Cinchona, with large, dense panicles of large flowers of a delicate pale pink, and very fragrant. Our hot-house and greenhouse are flourishing.

With much love to all your family party, and every wish that 1880 may be a happy year for you,
Believe me ever your loving brother,
CHARLES J. F. BUNBURY.

JOURNAL.

December 30th.

Exit the year 1879.

God grant that the coming year may open more cheerful prospects, and bring peace and plenty to all the world.

Eminent or remarkable persons who have died in the course of 1879, *with whom I was not personally acquainted :—*

Louis Napoleon, the Prince Imperial; Marshal Espartero; Marshall von Roon; Lord Lawrence; Mr. Roebuck; E. M. Ward (the painter of the South Sea Bubble and many other well known works); Panizzi; Delane; Michael Chevallier; Viollet Leduc; Sir Rowland Hill; Sir John Shaw

Lefevre ; Sir Louis Cavagnari ; Lady Waldegrave ; 1879. Fechter and Buckstone, celebrated actors ; William Froude, famous for mechanical science, and especially for his researches and discoveries relating to the construction of ships.

1880.

JOURNAL.

January 1st.

I take my leave of the past year, as of former 1880. years, with feelings of deep and earnest gratitude to Almighty God for the blessings which have been granted to me during its course. It is true that the retrospect is not one of such unclouded sunshine as that of some other years :—as to health for one thing. I endured a good deal of bodily suffering— not perhaps actual pain, but irritation which at times was almost or quite as bad as pain—from the gout, or eczema, which attacked me at the beginning of June, and plagued me grievously the whole of that month and the next, and part of August ; but though some traces of it still linger on, it now troubles me very slightly, and in other respects my health is excellent. The bad season, which has been so disastrous to all concerned in agriculture, has, of course, not spared me, though we have not, as yet, suffered any considerable actual loss, we have had to sympathize with the distresses and alarms of the farmer, and to contemplate the

1880. probability of a serious, perhaps a permanent reduction of our means. I have found it necessary to deduct ten per cent. from my rents for the present year, and if (which may God avert) the harvest of 1880 should be nearly as bad as that of 1879, the eventual diminution of our resources may be much more. But we have nothing for it but to hope for better things, and to trust that God will enable us to bear either fortune.

The past year has been a sad one to many, and an anxious one to very many in another way. Many valuable lives have been sacrificed in the Zulu and the Afghan wars. We have not indeed had any personal friend actually engaged as yet in these struggles : but we were deeply grieved and shocked by the death of Captain Ronald Campbell, a brother of our dear young friend Lady Muriel Boyle ; and in other cases also, friends of our friends have been exposed to imminent danger. We may now hope that the South African affairs are settled for some time to come—if not permanently — and that we shall not be troubled (perhaps not in my life-time) with any more Caffer or Zulu wars. The case is otherwise with regard to Afghanistan ; the prospect *there* is much more cloudy and uncertain, if not actually threatening, and I fear there will yet be, from time to time, much more fighting and many valuable lives lost before anything like settled peace is established in that quarter.

To return to ourselves : it is a matter of great and just thankfulness, that in the past year we have lost no one very near and dear; indeed no one 1880. whom I should call an intimate friend, except dear old Mr. Mills, and to him, from his great age, his blindness and infirmity, death must have been welcome. Mr. W. G. Clark, of Trinity, could not be counted as another exception, for he had been in effect dead (through the loss of his mental faculties) for several years. Stop: I have forgotten Sir William Boxall, another old friend. Canon Beadon, the *centenarian*, we had met several times at Wells, but could not rank him as more than an acquaintance. I had some acquaintance also with John Miers, the botanist, the author of several valuable papers on the botany of South America.

I ought not to have omitted to mention among the deaths, that of poor William Winthrop, a first cousin of Fanny, but one whom we had not seen for years, though formerly at times we used to meet him pretty often. He seemed to me a good-natured, amiable man, without energy, without purpose or definite occupation, but I really did not know him well enough to judge him fairly.

————

January 8th.

We had a very pleasant visit, in the latter part of last week and beginning of this, from Augustus and Rachel Vernon Harcourt and Dudley Hervey. Rachel is the second daughter of Aberdare (Henry Bruce) by his first wife, and so, a step-daughter of my dear cousin Norah; she is very handsome, very well educated and cultivated, with remarkably pleasant manners.—In short, she is charming.

1880. Her husband, Augustus Vernon Harcourt, is a professor or lecturer (on chemistry, I think), or perhaps on some particular branch of it at Oxford ; I like him much : he has nothing of the College Don about him, but has very pleasant manners, quite free from pomposity and dogmatism. He is lively and cheerful, and at the same time very well informed. In fact, he really seems to be worthy of such a wife.

Very good talk among all the party—lively and cheerful, yet far from frivolous.

Dudley Hervey I have mentioned before, though I do not well remember in what year it was that he visited us. He is now again in England on sick leave from Singapore, where he has been (if I am not mistaken) a magistrate. He is a very agreeable young man, with the true Hervey charm of manner, with a great deal of knowledge and mental activity.

He has, I believe made a particular study of the eastern languages, and has much knowledge of the ethnology of those countries.

Without being profoundly scientific, he has a great taste for natural history, especially botany, and appears to me to be a very good observer. He has very kindly made for me, a large and fine collection of the ferns (including also a few species of flowering plants, particularly fine specimens of a Nepenthes) of Singapore and Penang, which he brought hither, and which we looked through together. It is very interesting.

My nephew Harry also joined the party, and played well his part in it.

Our dear Arthur MacMurdo arrived from his tour safe and well, God be thanked—and looking bright and happy. He seems to have enjoyed his rapid tour immensely; his letters to Fanny from the Riviera, Genoa, Florence, Rome and Naples, were excellent, superior to what I thought him capable of writing; they show much intelligence and power of observation; and, hasty as this journey has been, I think it will have left impressions which will permanently benefit his intellect.

January 23rd.

Sir Henry Blake, my cousin, died yesterday morning early, from exhaustion, following on a severe attack of bronchitis. He was 86 years old, or at least in his 86th year. I have known him ever since the year 1822 or 1823; and in former days when I lived unmarried at Barton, and he and his first wife lived with his brother William in the Vicarage here, I saw him constantly and knew him intimately. He was a good man, of no brilliancy and no particular talents, but with a good deal of strong, rough, practical common sense; with little cultivation (beyond what could be gained at an ordinary school); kind and good natured, without delicacy of feeling, in short, "*abnormis sapiens, crasâque Minervâ.*" He has seen little of the world, either in the geographical or social sense; his views were narrow and his prejudices strong. But he was a good, honest, upright, kind-hearted man; a

1880. sincere and steady friend, and excellent in all his family relations. Henry Blake received his school education at Midhurst, where he was a contemporary and acquaintance of Charles Lyell.

His first wife (with whom my father and brothers and I were very intimate), was a very pretty woman, clever, lively, superficially well read, sparkling and agreeable, with a considerable turn for satire.

Without being as eccentric as his brother Patrick, he had many little amusing oddities and peculiarities for which he was quite willing to be laughed at. For one:—he had travelled very little—scarcely at all, and he always professed complete incredulity as to all remarkable and extraordinary facts which were told him, as observed in foreign countries —even when the observers were his intimate friends. We used to say (and he did not deny it) that he was determined not to believe anything that was different from what he had himself seen in Suffolk or Sussex.

———

January 28th.

We had last week (from the 21st to the 26th) the very great pleasure of a visit from dear Kate Hoare. She is looking very well, and is as charming and loveable as ever—I cannot say more.

Besides her and her husband, we had at the same time (for the Bury ball) Cissy with Emmy and Harry, the Barnardistons, with two of their daughters, Mr. and Mrs. Sancroft Holmes, Dudley Hervey and Captain Crosbie.

The weather has again become very severe. The first two days of this month were very mild, almost warm, the thermometer not going lower than 47 deg. on the 1st, and 45 deg. on the 2nd: the 3rd was a beautiful day, with slight frost in the morning, and for several days after (though it froze almost every night), the thermometer did not go *much* below the freezing point. But in the night of the 17th to 18th, it went down to 20 deg., and from that to the 30th, the nights were intensely cold, and the days not much warmer except in the sun. In the night of the 27th and 28th the thermometer was as low as 13 deg. (the greatest cold we have yet had this year), with a very thick fog, which cleared off in the afternoon, leaving the trees enveloped in most beautiful feathery coats of dazzling ice-crystals. The same fog and the same beautiful rime-frost prevailed on the 29th, when the frost lasted the whole day. The two last days of the month have been beautifully fine, clear and still, with moderate frost. All through the month the weather has been remarkably dry.

February 1st.

A beautiful day, bright and sunshine, actually *warm* in the middle of the day—moderate frost.

Saw the first symptom of spring:—a Crimean snowdrop, not yet open, but showing its flower-bud and the tips of its leaves above ground: and some

1880. winter aconites (Eranthis hyemalis), just beginning to peep above the surface.

LETTER.

My Dear Katharine,

Very many thanks for your agreeable letter. I am exceedingly glad to hear of Frank's accession of fortune, of which I have no doubt he will make a very good use. As Leonard is already well provided for, it is thoroughly satisfactory that this should fall to Frank. I am not sure that I ever before heard of Mrs. Sillery, but I respect her for making such a judicious disposal of her fortune.

By-and-bye, we may imagine Frank and Emily walking by the side of the "murmuring river Swale," like Wordsworth's *Peter Bell*, though they will certainly not be like him in other ways. I have an idea of those Yorkshire *dales* as very romantic in their scenery, as well as their associations.

I hope your health has not suffered from the terrible fogs which seem to have prevailed in London. It is a great mercy that Fanny and I are as yet quite well, but there has been a great outbreak—almost an epidemic, of throat complaints, bronchitis, and lesser forms of illness—among our servants, and in the parish generally. It is attributed to the frost-fogs of the 28th and 29th. Poor Allan, the gardener, has been alarmingly ill,

but I trust he is recovering. I cannot help doubting 1880.
the soundness of Kingsley's theory as to the
special healthiness of severe frosty weather. The
last three days, however, have been beautiful, with
bright and really *warm* sunshine, and no excessive
frost. Yesterday I saw the first snowdrop and the
first winter aconite of the year, at least the first that
I have seen, and welcomed them joyfully.

Dudley Hervey, who is in England on sick leave,
has brought me a large collection, which he had
very kindly made for me, of ferns from Singapore
and Penang; several of them very interesting,
especially Dipteris (or Polypodium) Horsfieldii.
Many of them are the same as yours from the
Khasia mountains, and others I had had from
Joseph Hooker; I am not quite sure whether any
are quite new to my collection, but they are very
interesting altogether, and very fine specimens : and
of most of them there are numerous duplicates. I
will write more particulars to you when I have made
more progress in arranging them.

Have you looked into Edward's book? I have
read the greatest part of the first volume, and am
very much pleased with it; it is a wonderful piece
of industry and care and learning, and is moreover
very clearly and pleasantly written. I am reading
also the new volume of Lord Minto, containing his
Governor-generalship of India; in this there is
much that is interesting, but I am very much struck
with the contrast between the pleasant, lively style
of his private letters, and the stiffness and *verbiage*
of his official correspondence.

1880. I have been much interested by two articles in the January and February numbers of *The Cornhill*, by Dr. Hunter, on "What the English have done for the Indian People." He seems to make out a very good case for us, but he promises another article on what still ought to be done.

I like some of Miss Cobbe's "*Re-Echoes*" very much.

I hope Leonard will be successful in his election, though I am afraid his political opinions do not agree with mine, but I am sure he is at least an honest politician. It is indeed a shame that elections are so costly.

I have just heard that our doctor, Dr. Macnab, is ill in bed; there seems to be a regular influenza. I cannot be too thankful that Fanny and I are well, and Arthur also. I hope you are.

Believe me ever,

Your loving brother,

CHARLES J. F. BUNBURY.

JOURNAL.

February 2nd.

Another very fine day. Saw one or two more Snow-drops.

February 4th.

My 71st birthday. I can only repeat what I wrote a year ago; that I feel I owe most profound and humble gratitude to the Almighty, for the undeserved goodness He has shown me in so many 1880. instances all through my life and now in allowing me to reach this age in health and wealth, surrounded with so many blessings. Above all, I have cause to be grateful more than I can express, that my inestimable wife is preserved to me in good health, and that our union of love is as firm as ever ; also that we still enjoy so many dear and valued friends, as well of the younger as of our own generation.

LETTER.

Barton, Bury St. Edmund's,
February 5th, '80.

My dear Katharine,

Thank you, most heartily, for your very kind letter, and for all your good wishes and kind expressions ; also for the very pretty book you have sent me.

We are pitying you who are plunged in the terrible fogs of London ; for I see in the papers that that there was a very bad fog yesterday, while here it was one of the finest winter days I ever saw, or rather indeed a very fine spring day.

You shall be very welcome to plenty of duplicates of my Malayan Ferns, as soon as I have got them into tolerable order, but I have some doubts whether any of the species will be new to you. However, the localities—Penang and Singapore, may be new and you may be glad also to have additional specimens to exchange with other collectors.

1880.   Dear Arthur left us this morning in very good
spirits, to go to his new tutor, Mr. Hiley, near
Buntingford in Hertfordshire, which seems to be
now rather an out of the way place, though
formerly it was on the great north road. I hope
it will agree with him, and that he will work well.
He danced all night long till past six in the
morning, at the servants' dance on the eve of my
birthday.

Ever your loving brother,

CHARLES J. F. BUNBURY.

JOURNAL.

February 13th.

Scott tells me that the wheat is at last appearing, after having been so long kept dormant by
the long-lasting frosts of this winter, and that it
looks well, not appearing to have been injured.

There has been a terrible mortality, it appears,
among the sheep in many parts of England; the
disease called the *rot* having prevailed among them
to a frightful degree; owing no doubt to the
excessive wet through the spring and summer of
last year. The disease (Scott tells me) has been
very bad in the fenny country about Yarmouth, but
he has not heard of its yet showing itself in our
part of Suffolk.

LETTER.

Barton, Bury St. Edmund's.

February 19th, 1880.

My Dear Edward,

I was extremely glad to read the letter which Fanny received from you yesterday, as I had become anxious lest you should have been ill, as so many people have been, from the dreadful fogs which seem to have prevailed in London. It was a great comfort to read such a good account of your health, and it is really a great mercy that Fanny and I have continued quite well (hitherto) through all this winter.

Poor Lord Hanmer whom you mention is in great distress; his wife is dying. Fanny had a letter from him yesterday, in which he mentioned that he had bought your book; for, he says, the subject is one in which he has always felt a great interest, and he knows that no one is so well qualified to treat it as you.

We have lost our excellent old servant and friend, Mrs. Marr, at Mildenhall ; she died the day before yesterday, quietly and easily, without pain, we hope. She was 81 years of age, and had for some years past been very infirm, and very deaf; but Fanny feels much the parting with such an old friend ; for Mrs. Marr had been in Mrs. Horner's service for 15 years, and since in ours for very nearly 30 years, connected with Fanny's family for altogether 45 years. She was an excellent woman, and intelligent and cultivated to a degree quite unusual in her rank of life.

1880.   Poor Lady Blake, I hear, wishes to live in Bury.

I am very glad to hear your book has such a success, which I am sure it well deserves. I have finished the first volume, and continue to like it very much, though I find part of the section on Eratosthenes, rather stiff, but this is because I am rather ignorant of the *geometrical* part of geography.

I have read the new volume of Lord Minto with great pleasure, and actually wished that it had been longer—not that there had been more dispatches, but that there had been more of Lord Minto's private and familiar letters, which are exceedingly pleasant reading; he seems certainly, to have had a special talent for that kind of writing. This volume too, as well as the former one, gives me a very favourable impression of his character.

Fanny is delighted with the volume she has read of McCarthy, and I like very much the portions she has read to me. I have been reading to her in the evenings, Sir George Lawrence's "Narrative of his own Experiences in the First Afghan War;" he was one of the captives. It is deeply and painfully interesting. Old as the story now is, one cannot help being almost miserable as one reads the story of such deplorable imbecility in the chiefs, and such dreadful sufferings of their followers. Our present Afghan war was not, I think, much wiser or more just in its beginning, but it has certainly been carried on hitherto, under very different sort of commanders.

We have had very high winds now for several days, but a very mild temperature, and the young corn, Scott tells me, is looking as well as possible.

We have plenty of snowdrops in the garden, and the 1880.
crocuses are beginning.

Ever your very affectionate brother,

CHARLES J. F. BUNBURY.

JOURNAL.

February 24th.

Talking of Female Suffrage, Mr. Lecky said (very rightly I think), that if there were any reason to believe that a large proportion of educated and respectable women wished earnestly for it, he thought that it ought to be granted, and indeed that in *that* case it certainly would be granted before long; but he did not believe that the fact was so. I agree with him; at least I certainly think that before such a measure is carried, pains should be taken to ascertain the opinions of the better sort of women. He said it seemed hard to refuse married women the suffrage if it were given to maidens and widows, but he thought the giving it to the married would only increase the amount of matrimonial dissension and misery, and in the worse portions of society add to the ill-treatment of wives. Experience, he said, has clearly shown that the ballot was quite ineffectual against priestly tyranny, and he believed it would be equally so against marital tyranny.

February 25th.

Mr. Lecky said that the agitation for Women's

1880. Rights in these latter days had really had the effect of directing public attention to the real grievances of women, and had led to the redressing, more or less, of some of them (as for instance, in the law protecting a married woman's earnings from a bad husband); and he thinks that all that is really important will by degrees be gained in this way. It is thus (he said) that most of our reforms have been effected :— by demanding at first something more than the reformers expected to gain, and so gaining the essential by a compromise. He illustrated this by a story of O'Connell :—Somebody asked O'Connell whether he really expected to gain the Repeal of the Union. He answered, "No, I don't think I will gain it; but I think that by agitating for it, I will gain all that we want besides."

Mr. Clements Markham showed as a very curious old map of the year 1600, seemingly the very one which Shakespeare speaks of (in *Twelfth Night*) as "the new map with the augmentation of the Indies;" it is crossed over with an inconceivable number of lines in every possible direction, giving it a very strange appearance. It was copied (if I rightly understand) from a Dutch map which was published the year before, immediately after the Northern discovery of Barents. A vast number of names of places are marked in it (in very minute characters), but all on the coasts, none inland, showing that it was intended chiefly for the use of seamen. The outlines of Africa and South America, and of the eastern coast of North America are drawn with approximate correctness; and a little bit of Australia is introduced (but I think not 1880 named).

Mr. Markham said that this curious map (of 1600) was the first English one constructed on the principle called "Mercator's Projection;" Wright, by whom this map was drawn, had studied under Mercator.

Mr. Markham explained to me what I had not understood before, as to the voyage of the *Vega*, under Professor Nordenskiold;—that its importance consisted not in its having made any new discoveries on a large scale; the general outline of the northeastern extremity of Asia had been seen and mapped *from the land side* by Russian explorers, especially by Wrangel, several years ago; but no ship had reached or passed through Behring's Straits from the west, and they were supposed to be perpetually blocked up by ice.

Mr. Markham told me he was at the Geographical Society when Mr. Ball read his paper on the Flora of the Alps; and that all, even Hooker himself, seemed puzzled and *posed* by the speculative part of it, and as if they did not know what to say to it.

Mr. Markham spoke of the great difficulty of cultivating in our plant-houses the natives of high mountain countries within the Tropics; the Chinchonas in particular. Mr. Howard (the author of a celebrated and splendid work on those plants, cultivates some species with success, but it is by giving them a house entirely to themselves, and devoting special care to them exclusively. At Kew, the Chinchonas are drawn up into very tall, thin, weak,

1880. *stalky* plants, bringing neither leaves nor flowers to perfection.

Mr. Markham included the Melastomaceæ among the *mountain* plants of tropical countries which are difficult to cultivate; but this is hardly correct as to those of Brazil; the beautiful *Pleromas* of the neighbourhood of Rio grow at very moderate elevations, certainly below 1000 feet; one of them indeed (Pl. argenteum), descends almost to the sea level.

LETTER.

Barton, Bury St. Edmund's,
February 26th, '80.

My dear Katharine,

Very many thanks for your kind letter. I know Mr. Bicheno's name well as an English botanist; he is often mentioned by Smith, Turner, and their contemporaries, and wrote a good paper in the *Linnean Transactions* on the British species of Juncus, and another on Orchis militaris. He was for several years Secretary of the Linnean Society, and I find in the Obituary Notices of that Society, that he was born at Newbury in Berkshire, the son of an eminent Baptist Minister :—that he lost the bulk of his fortune by an unfortunate investment in some iron works: that he was one of the Commissioners for inquiring as to the expediency of introducing Poor Laws into Ireland; that in 1842 he was appointed Colonial Secretary of Tasmania, 1880. and died in 1851 at Hobart Town.

I never wrote a paper for publication on the botany of the South of France, but I wrote a long letter on that subject to Charles Lyell, which I remember seemed to interest him. It was about the end of '47 or beginning of '48 : written, I rather think, from Nice.

The *Carubiere* is certainly the *Ceratonia Siligua*—*Caruba* being, I suppose, the name of the fruit, and Carubiere of the tree.

By the way, there is a very pretty sketch of the vegetation of the Riviera in Mr. Allman's Presidential Address, last year, to the Linnean.

I am very glad to hear that Caroline and Sophy Lyell are going on well. I was much grieved for the death of Marianne, though at her age it could not be quite unexpected. I can never forget the kindness of all the sisters to us during Fanny's terrible illness at Kinnordy, and at a time when they had so much anxiety and sorrow of their own.

Ever your loving brother,

CHARLES J. F. BUNBURY.

P.S.—I am not sure whether I have told you already, that two Weymouth Pines which were blown down in one of the groves here, in a great snow-storm, have been measured ; one was 80 feet high, the other 84. Both were planted by my father.

JOURNAL.

1880.   Mr. Markham has lent us to read, a M.S. paper
or pamphlet on the present Afghan war, which he
intends to publish, or at least to print for private
circulation.   I think it most excellent—particularly
clear.   It shows, with remarkable distinctness,
the different steps of the transactions between
our successive Governments and the Afghans ; the
broad difference between the policy pursued by the
Governments of Lord Lawrence, Lord Mayo and
Lord Northbrook, from 1845 to 1875, and the policy
of Lord Lytton, since 1875, and the contrasted
results of the two systems.

Mr. Markham spoke of Moseley's book ("The
Voyage of a Naturalist on the Challenger,") as a
delightful book, in which I quite agree with him.

On the 26th of February, Mrs. Markham showed
me a choice collection of plants from Nova Zembla
(Novaya Zemlia).   They are of course of the same
general character as those from our Arctic expedition
which she showed me last February, and many or
most of the species are the same ; but what is
curious there are also a few common English
plants, such as our common Dandelion and Carda-
mine pratensis.   Some of the most interesting I
noticed were—Saxifraga cernua (many specimens,
some much taller than any I had seen before, and
with much of the habit of S. granulata).   Saxifr.

cæspitosa, oppositifolia, and flagellaris; Polemonium 1880.
cæruleum (very dwarfish, with large and beautiful
blue flowers) ; Salix arctica and polaris.

March 3rd.

A remarkably agreeable party here since Feb-
ruary 21st. Lady Muriel and Mr. Courtenay Boyle,
Lady Grey, Mr. and Mrs. Lecky, Mr. Clements
Markham, Mrs. and Miss Markham, Lord and Lady
Rayleigh, Miss Balfour, Lord John Hervey, Dudley
Hervey and Edward. Not all at once however ; for
the Markhams arrived on the 24th, the Leckys and
the party from Terling went away the next day; and
Edward did not come till the 28th.

Mr. Lecky and Mr. Clements Markham are two
particularly agreeable men ; I hardly know any
more so ; they are very unlike yet I can hardly say
which is the more agreeable. Mr. Lecky I did not
know well before this visit, and was very glad to
improve my acquaintance with him. He is full of
knowledge, and ready to communicate it in a very
agreeable manner. History (especially the history
of manners and opinions), and literature, seem to be
his special topics. Markham appears to be also
very strong in history, though the subjects for which
he is most celebrated are scientific.

March 8th.

Scott, who returned on the 6th from Mildenhall
where he had been holding the rent audit for

1880. me, gave a pretty good account of his success. I have given him instructions to abate 10 per cent. of the Mildenhall rents, but nevertheless he paid into my account a balance of £1800 (only £300 less than had been usual). He says, indeed, that most of the farmers about Mildenhall and that part of the country are much impoverished and distressed, especially those in the fens, where the wretched season and bad harvest were felt with particular severity. But what is a comfort, is he says that none of the tenants complained of their rents, or said anything against me, but on the contrary, expressed themselves as grateful to me.

March 9th.

We learned from the Ipswich morning paper, that Parliament is to be dissolved almost immediately. Rather a surprise.

March 13th.

Fanny made me a present of a beautiful book— Emerson's "Trees and Shrubs of the Massachusetts," the illustrated edition (of 1878). The uncoloured plates are very clear and accurate,— the coloured ones exquisite, particularly those showing the autumnal tints of leaves. She had intended it to be a birthday present to me, but it did not arrive in time.

March 14th.

A pleasant party here from the 9th to yesterday;

—Miss North, Mr. and Mrs. Hutchings, Mr.
Walrond, Agnes Wilson; Mr. and Mrs. Ferguson
Davie, and Mr. Bowyer.

Miss North is an extraordinary person. I have
already more than once mentioned her wonderful
flower paintings. Though by no means young, she
is now just about to start for Sarawak, Borneo, to
spend some considerable time there, to continue and
complete, as far as she can, her series of pictures of
Malayan flowers and fruit; thence to go on to
the Australian Colonies and New Zealand, and to
visit the Cape in her way back. She travels quite
alone, without even a maid; goes into the wildest
parts of the most unhealthy and dangerous countries
—and seems equally free from fear and from harm.
Yet there is nothing bold, masculine or eccentric,
either in her appearance or manners; on the contrary she is very quiet, gentle, and unassuming—not
a great talker, though willing to communicate her
knowledge; in short, a very pleasant person as well
as a very remarkable one.

Agnes Wilson is charming; Mr. Walrond very
agreeable.

———

March 17th.

The newspapers mention the death of Thomas
Bell, the author of the excellent books on British
quadrupeds and reptiles,—a very good and useful
naturalist, and a very good-natured worthy man.
He has lived to a great age, 87, according to the
newspapers. He had bought Gilbert White's house

1880. at Selborne, and spent there all the latter years of life, and his edition of "White's Selborne," is a very satisfactory book.

————

March 25th.

We have heard within very few days of the death of two of our friends — Lady Augusta Seymour, and Lady Hanmer. The first was the elder sister of the Bishop of Bath and Wells ; much older than him, for she was (according to the newspapers) in her 81st year ; so that her death could not have been entirely unexpected. Yet I know that it will have been a great sorrow and a grievous loss to him; not merely because she was his last surviving sister, but because I know that there was a very warm affection and attachment between them.

Lady Augusta must have been very handsome when young ; her niece, our dear Sarah Hervey, had a striking likeness to her. For myself, I knew but little of her until we met her at the Bishop's Palace at Wells, in September, 1871, and spent some days in the same house with her ; she is thus inseparably associated in my mind with the memories of that delightful time, and of course my impression of her is the more favourable. But allowing for this influence, she appeared to me a very interesting and very agreeable woman, and there seemed to be something congenial between us. She was one of the party in a delightful afternoon's excursion, never to be forgotten, when we drank tea in a lovely little valley on the slope of the

Mendips near the village of West Horrington: 1880.
dear *Sarah*, Kate, their brother James, Fanny and I,
making up the party. Since then, we have now
and then seen Lady Augusta in London, and she
has always been very cordial and friendly; we have
repeatedly *tried* to meet, but it is very difficult in
London to meet those we care for.

Lady Hanmer and her husband were devoted to
each other, and he, poor man, is truly to be pitied
in his loneliness. He wrote to Fanny to tell her of
the termination of the long suffering.

March 31st.

We heard of the death of our friend Mrs.
Abraham—which took place suddenly—at Florence.
Many years ago, when we lived at Mildenhall, and
her husband had the living of Barton Mills, they
were our nearest neighbours, and a great intimacy
and friendship grew up between us. She was then
very pretty, with very gentle, pleasing manners, a
refined and attractive nature,—altogether a very
captivating person, with a charming talent for
music and a cultivated and graceful mind.

======

LETTER.

Barton, Bury St. Edmund's,
April 3rd, 1880.

My dear Edward,

The elections are really becoming rather
exciting; at the beginning, I did not care much

1880. about them, I did not suppose they would make much difference : but now it seems as if the tide were really setting strongly in favour of the "Liberals," and likely to carry them into power again. Were you not surprised at Mr. Hardcastle's success?—so great a majority?—I was, and so were most of those whose opinions I have heard. I do not know what Mr. Hardcastle and his particular friends expected, but I think the general opinion was that it would be a very close contest. I hear, however, that Lord Francis had been very remiss in his attentions to the voters: and I know of old that they are very sensitive on that score. Since I began, I see to-day's *Times* says, it is now *certain* that the present Ministry will very soon be replaced by a Liberal one. I must say I am not glad of it.

I concur entirely with Lord Grey—whose letter I heartily approve and admire:—I think Mr. Gladstone a very dangerous man, and I think it quite certain that *he* will be Prime Minister in fact, whoever may be so in name. Nor do I like the prospect of the Government being carried on upon a see-saw principle—that every general election is to be followed by a total change of Ministry and of policy. No doubt the Home policy is more important than the foreign, but I cannot think that other nations will look upon us with much respect or confidence if they see our policy entirely reversed every seven years.

Dear Sarah and her two little boys are with us now—a great pleasure. Charlie is a delightful little fellow.

I hope you have kept in good health. Fanny and 1880.
I have, in turns, had severe colds, but I am thankful
to say we are both well again.

We have received accounts of the death of our
friend Mrs. Abraham, at Florence : it seems she
caught a cold which attacked her heart, which was
already in a diseased or weak condition, and her
death was almost sudden. It is not many years
since she was a very attractive and interesting
person, as well as very pretty.

I have not yet finished your book, which is
certainly a *piéce de résistance:* but I continue to
admire it, and to find much that is curious and
interesting in it, — particularly latterly in the
accounts of India. Besides, it has the good effect of
making me rub up my own geographical knowledge.

Do you know that a true Cedar (but more closely
resembling that of the Atlas than of Lebanon), has
been discovered on the mountains of *Cyprus* by Sir
Samuel Baker, who sent specimens to Hooker.

With Fanny's love, believe me ever,

Your affectionate brother,

CHARLES J. F. BUNBURY.

JOURNAL.

April 5th.

The elections have taken an unexpected and surprising, and (to me) unsatisfactory turn. They are
not, indeed, yet concluded ; but they have gone
so far as to make it evident that the "Liberals"
(including the "Home Rulers"), will have a decided

1880. majority in the new Parliament. The *Times* of last
Saturday, indeed, speaks of it as "now *certain*"
that the present ministry will very soon give place to
one of opposite opinions; and the *Globe* has little to
say on the other side. What is the reason of this
startling change since 1874? I cannot think that
the majority—the common run—of voters care
enough about foreign politics to have been seriously
influenced by the Eastern, or Afghan, or Zulu
questions; and I am not aware that there has been
of late years any domestic question of exciting
interest, like those of the beer traffic and of secular
education in 1874. I am afraid it is the effect of
fickleness and love of change in the mass of voters.
It will be very unfortunate if our country should
come to be governed in a *sea-saw* fashion—a complete change of policy with every general election.
This is what the experience of 1874 and '80 seems
to indicate. For my part, I am by no means delighted
with the prospect of an immediate change of ministry and of measures. I go entirely along with
Lord Grey, whose letter (published in the *Times*
a few days ago), on the Northumberland election, I
thoroughly approve and admire. I think Gladstone
a dangerous man (though I have no doubt he is
an honest and sincere one), and all the more dangerous because of his splendid talents; and I have
no doubt that, even if he refuses all office whatever
he will be the ruling spirit of the party. Lord
Hartington and Lord Granville would be in themselves unobjectionable).

The election for Bury was the first surprise—not

only to me, but to those with whom I had spoken on the subject. Scott, whom I think a very good judge of such matters, thought that both Conservatives would probably come in, but that it would be a close contest; every one indeed seemed to believe that it would be a very "near thing,"—none to expect a large majority such as Mr. Hardcastle actually had.

April 15th.

The General Election is now nearly over, and the result is a still more complete victory of the Gladstonites—a still more overwhelming defeat of the Conservatives than seemed to be threatened ten days ago. Gladstone's victory in Midlothian is itself an astounding triumph.

I look forward with great misgivings, great anxiety—not to say fear—to the doings of the new Parliament.

We had, two days ago, the sad news of the death of our friend Lord Charles Hervey:—sad but not unexpected, for we heard a few days ago, from his daughter Isabel and his son Dudley, that he was very dangerously ill, and indeed that there was scarcely any hope. Inflammation of the lungs I believe was the cause of his death. I regret him very much; I had a great liking and regard for him, and always took great pleasure in his society,—especially in hearing him talk of the many countries he had visited, and of their natural history.

The doctors (Dr. Andrew Clark in particular), urgently advised him to go to a warmer climate for

1880. the winter, as he had usually done: but he thought
it his duty to try to continue in charge of his
parish through the winter, and the result has been
fatal. I grieve for all Lord Charles's family, and I
grieve especially for our dear Bishop, Lord Arthur,
who is now left alone of his family in that
generation, having survived five brothers and three
sisters. It is very sad; I know his warm heart and
strong family affections, and am sure that he will
feel it keenly.

April 16th.

I was sorry to see in the newspaper the other
day, the death of Professor Schimper, of Strasbourg.
He was a very eminent botanist, and his works,
both on Mosses and on Fossil Plants, are of great
value. I possess his "Synopsis Muscorum Euro-
pæorum," and "Traité de Paléontologie Vegetale,"
and study them very often: they are excellent. I
made acquaintance with him when we were at
Strasbourg, in 1848, and found him very kind and
courteous, and remarkably agreeable. He had very
pleasing manners, and a highly cultivated mind:
had travelled extensively, zealously exploring, as a
botanist, nearly every country of Europe, from the
south of Spain to the extreme north of Lapland:
and his conversation was both lively and rich in
matter.

M. Schimper, being an Alsacien, spoke both
French and German, and told us that both were
equally easy to him. He spoke French, indeed,
evidently with perfect ease, and no doubt with

perfect command of the language, but with a 1880.
decided German accent. I understand that, after
the war of 1870, he *elected* to be a German, and
continued to live in Strasbourg.

———

It is announced that Gladstone has been offered
the post of Prime Minister (with that of Chancellor
of the Exchequer), and has accepted it. So now
we are *in for it*. I *wish* that things may turn out
better than I *expect*.

———

We parted (very sorry to part), with dear little
Charlie Seymour. His dear mother, Sarah,
brought him and his little brother here on the 31st
of March, and stayed with us till the 16th of this
month, when she went to rejoin her husband at
Southsea, leaving the little ones with us: but they
are now claimed by their grandmother, Lady
Hertford. It is always a great pleasure to have
Sarah with us.

———

The new Ministry is now formed, and a queer
medley it appears to be,—some of the members
very good and some very bad. How will Lord
Selborne and Lord Hartingdon work with Mr.
Bright and Mr. Chamberlain and Sir Charles
Dilke ?

1880.   May 1st.

A beautifully bright, sunny day, unlike many
May days that we have seen of late years,—a sharp
east wind, but the brilliant and really powerful
sunshine very exhilarating.

May 3rd.

A sorrowful farewell to dear Minnie, who left us
to go to London on her way to Southsea. I am
always sorry to part with her.

LETTER.

Barton, Bury St. Edmund's.
May 3rd, 1880.

My dear Edward,

We propose to go to London on or about
the 20th of this month. Everything here is in such
beauty, the season is such a fine one, that I must
confess I do not at all like the thought of leaving
the country; but we have been completely stationary
for ten months, and I suppose it is desirable to
do something to rub off a little of our rust and know
a little more of what is going on in what is called
the world; though year after year I feel the London
world less and less congenial to me.

The new Ministry appears to me a mere medley;
how long will Lord Selborne and Lord Hartington
go on comfortably with Mr. Bright and Mr. Chamberlain? I do not think it can be long before
an explosion will take place in the Ministry, and

I fear in that case the violent party will win, and 1880.
there will then be little safety either for the Church
or for the English system of landed property. I
hear there is a report that Mr. Lowe is to be
a Peer.

We have been much alarmed for George, as he is
with General Ross's division, and must have been in
the last action ; but the accounts from Caubul
in to-day's *Times* have put us more at ease, still
there must always be danger among such a ferocious
and lawless race as the Afghauns.

Leonard and Mary Lyell are here at present with
their three very nice little children. Minnie left
us this morning.

I shall very soon be qualified for a place in the
Order of *Edentata*.

Ever your affectionate brother,

CHARLES J. F. BUNBURY.

JOURNAL.

May 4th.

The Leonard Lyells went away, having been here
since the 27th of April. Mary Lyell is a very
clever, well informed, pleasant-mannered, agreeable
woman,—very active-minded. She is very eager
about politics, and I think has inspired her husband
with like zeal.

Their three children, Charles, Nora and Nellie,
are very nice.

1880.   May 5th.

While Minnie was here alone with us, I read
aloud to her and Fanny in the evenings, Sir Henry
Taylor's "Philip van Artevelde." Fanny indeed
had read it before, and so had I, but long ago. We
were all delighted with it.

_____

May 10th,

I fear there can now be hardly a doubt that the
Atalanta has gone down with all on board.
Scarcely anyone appears to retain a hope. It is a
terrible and deplorable misfortune.

_____

May 12th.

Mr. D. T. Fish* came over from Hardwick, and
we went through the plant-houses with him. He
seemed very much pleased with what he saw, and
expressed warm admiration of the houses (built by
Weekes, of the King's Road, five years ago), and of
many of the plants. He noticed particularly the
Quisqualis Indica (which has indeed grown finely,
and is flowering profusely),—Xylophylla angustifolia,
Brunfelsia Americana (which he had never seen in
flower before), the Anthurium Scherzerianum with
an uncommonly large spike of fruit, and our ferns,
especially the great Alsophila tenerum var.
Farleyense. Also gave us several useful hints
for the management of others. I was surprised to
hear him say that the hardy out-door plants and
trees are more forward, more developed here than

* For many years Lady Cullum's gardener.

at Hardwick. I have observed this year, that the 1880. earliest flowering plants—the Yellow Aconite, Snow-drops and Crocuses—were quite a month later than in ordinary years, but that, as the season has advanced, vegetation has been less and less behind-hand : and now the Horse-chesnut and other flowering trees appear to be quite as forward as usual.

———

May 14th

Most beautiful weather. I delight in the spring weather; old as I am, I do not think that my enjoyment of nature is at all deadened. I feel the "vernal delight" keenly,—the beauty of the flowers, both in the gardens and in the fields, the brilliant freshness of the grass, the exquisite delicacy and beautiful variety of the colours of the young leaves, the liveliness of the birds, the first notes of the cuckoo, the first sight of the swallow, all give me hearty enjoyment, and I feel very unwilling to change them for anything that London can give. It is true that with all these pleasures of spring, comes now and then a feeling of longing regret,— "*Saudades*," as the Portuguese call it.—"I turn from all she brought to those she could not bring." But it is a chastened feeling, and the beauty of spring helps one to receive the hope which religion inspires.

The Lilacs and Horse-chesnuts are in beautiful bloom, especially the large Horse-chesnut tree opposite the N. E. front of the house, which is our

1880. finest tree of the kind, and is always the first in leaf and in flower.   The Laburnums are rather late.

May 19th.

We went yesterday (in spite of a bitter east wind) to Hardwick, and went through the gardens and plant-houses with Mr. Fish, who was very pleasant. All in beautiful order and condition; it could not have been better if our dear old friend had been there. It seemed difficult to believe really that she was gone never to return.

We saw many fine and interesting plants in the hothouses: but the Pinetum is the glory of Hardwick; the Conifers there are really splendid specimens, and they are arranged so as to produce the best general effect.

May 22nd.

Arabella Bevan, the eldest of the remaining daughters of poor James Johnstone Bevan, died last night.   She has been long ill, almost hopelessly so, with consumption, and was kept alive winter after winter, only by removal to the south of England. She has been carried off at last rather suddenly, this last attack having come on a very few days ago. She was a very interesting, pretty, attractive girl when she was in tolerable health; but it is a good while since I have seen her.   I feel very great pity for her sisters, and above all for her poor father, who has already had the misfortune of losing his wife and one daughter before this one.   It is most sad.

We went up to London. Found dear Minnie established in 48, Eaton Place. Susan MacMurdo, Cissy, Emmie, and Harry, came to see us.

The weather lately has been delightful; brilliant sunshine, soft, mild air, warmth delicious and not oppressive. In fact, nearly the whole of this month has been splendid weather; bright, exhilarating sunshine, even when there has been a cold wind. I hardly remember to have seen Barton in greater beauty of its spring dress. I am very glad we have stayed late enough to see it in all the glory of its lilacs, laburnums, horse-chesnuts, hawthorns, white and crimson, moutan pæonies,—to say nothing of irises and multitudes of beautiful herbaceous plants.

May 30th.

Our 36th wedding day. The anniversary of a memorable day, full of blessings and happy memories for me. I can never feel grateful enough to the Almighty for his goodness in directing my choice in the most important crisis of my life, as well as in permitting me so long to enjoy the blessings resulting from it.

May 31st.

We have heard of the death (sudden, if I am not mistaken, or nearly so) of Mrs. Gurdon (Henrietta Colborne), one of the daughters of my father's very old friend, Lord Colborne. She was an excellent woman.

1880.   Mrs. Gurdon was 70 years of age, one year younger than me.

———

Received a delightful letter from Rose Kingsley; Fanny had had an equally good one from her a few days before. She and her mother were in great spirits at having found a house and home in the neighbourhood which she describes as almost perfection — very near to Wormleighton, the parish where her brother-in-law's living is; and near also to the farm of which her brother Grenville has become tenant. In short, she describes it altogether as almost *fabulously* delightful, the only drawback being that the house will not be ready for them till the autumn. The name of the place is Tachbrook, and it formerly belonged to Walter Savage Landor.

———

I had a charming visit from dear Kate Hoare. She was looking well; and told me that she had seen Mrs. Kingsley in the winter (when she, Kate, and her husband were staying at Ladbroke, near Leamington), and had been astonished to see her looking so well — better indeed than when she had last seen her at Eversley; — and conversing with much of her old animation and agreeableness

———

Our dinner party of 19, including our 3 selves; very

pleasant. The MacMurdos (3 including Mimi), Gen-
eral and Mrs. Lynedoch Gardiner and Miss Gardiner
Mr. and Mrs. Gambier Parry and Miss Parry; Lady
Rayleigh; Katherine Lyell; Lord Hanmer; Sir
Edward Greathed; Sir George Young; Edward;
my nephew Harry. After dinner came in Mr.
Sanford and his two (very pretty) daughters. I was
very glad to see him; Ethel Sanford is quite lovely.

June 9th.

Dinner party of 18; Mr. and Lady Barbara
Yeatman, and Miss Yeatman, Kate Hoare, Harry
Bruce and his bride; Mr. and Mrs. Richard Strutt,
Mr. and Mrs. Drummond, Mr. and Mrs. Edward
Goodlake, Admiral Spencer, Arthur Hervey, Arthur
Lyell.

Harry Bruce gave a lively account of the Greek
play ("The Agamemnon,") at Oxford, from which
he was newly come; his half-brother Willy having
been one of the principal performers. The character of Cassandra (he said) was most admirably
performed and looked by Lawrence, son of the
author of Guy Livingstone.

June 10th.

Our dinner party—Bishop of Bath and Wells,
Lady Arthur, Caroline (*Truey*) Hervey. Lady
Edith Adeane, Mr. and Mrs. Lecky, Mr. and Mrs.
Walrond, Lady Mary Egerton, Leopold and Lady
Mary Powys, Mr. Goodlake, Edward, Harry,
Clement.

Our dinner party yesterday was a particularly agreeable one. The Bishop of Bath and Wells was there, and in great force, looking remarkably well and in very good spirits,—and he, Mr. Lecky, and Mr. Walrond, are three of the most agreeable men I know.

Mr. Lecky talked of Justin MacCarthy's History, of which his opinion agrees very well with mine ; said that MacCarthy is (in his Parliamentary character) one of the most wrong-headed and impracticable of the Home Rulers, and remarked how strange it is that he should have written his book with such striking fairness and impartiality. He knows Carlyle well, and says he is still in perfect possession of his faculties, and he (Mr. Lecky) often takes him out in his carriage. He agrees with me in especially admiring the History of the French Revolution and remarked that Carlyle has a great talent for drawing characters.

We went to see the Grosvenor Gallery, and specially to see Annie Pertz's picture. Very much pleased with it.

It is a small unostentatious picture ; the subject —" Old Caspar holding the skull which his grandchild has found on the field of Blenheim" (from Southey's fine ballad on " The Famous Victory.") The expression of the faces, and character of the figures, appear to me excellent—thoroughly appropriate, telling the story most clearly, without exaggeration ; the colouring natural and pleasing. It promises well for her future success in the art to which she has devoted herself.

I noted a few other pictures in the Grosvenor Gallery which pleased me ; especially one by *J. D. Linton*, entitled " Victorious." A General returned triumphant from an expedition, presenting himself to a King and Queen, who are seated in state to receive him. It is a composition of a great many figures, in the rich costumes and armour of the 16th century, and is painted with great richness and beauty of colouring. Also " The August Moon," by *Cecil Lawson* ; a large landscape, painted in a peculiar style, requiring to be seen from some distance, but, when so seen, having a fine effect.

June 14th.

We went to the Royal Academy ; a great crowd, but we managed to see some things.—

" An Assyrian Captive," by *Long*. Interesting, but not so much so as several of his previous works, as it is only a single figure ; but the expression is very good.

" Buonaparte on board the Bellerophon," *Orchardson*. A very fine, impressive picture.

" The Night Watch." *Briton Riviere*. Very striking ; lions and lionesses prowling in the clear moonlight, amidst the ruins of ancient temples. The effect of the moonlight, and the glaring eyes of the creatures are wonderfully given.

1880. Portrait of John Bright. *Millais.* Not so impressive as his Gladstone, last year, but the deficiency may be in the subject.

"From Hand to Mouth." *Faed.* The expressions of the principal figures—the poor man and his family—very pathetic, almost painfully so ; but there are some parts of the picture, of which the meaning does not seem well made out.

"Woodland and stream." *Charles Johnson.* This is a name new to me. It is a very beautiful, and beautifully painted scene, a richly-wooded, hilly country, evidently somewhere in the south or west of England—quite in the style of *Vicat Cole*, and I supposed it to be his till I consulted the catalogue.

We dined with the Charles Hoares. I was between Kate (who is always charming), and a Mrs. Godley, whom I found very agreeable.

June 15th.

We dined with Mr. and Mrs. Roundell; met Mr. Stopford Brooke (the preacher) and his sister, Mr. and Mrs. Garrett Anderson. I sat between Mrs. Roundell and Miss Brooke, a clever, highly cultivated, very agreeable woman. Mr. and Mrs. Roundell were very pleasant.

June 16th.

Dining with the MacMurdos, we met the Leopold Powyses, Lady Rayleigh (Dowager), Lord Mark Kerr, Lord and Lady Charlemont, and Mr. and Mrs. Moulton, Mr. Cardwell.

Mrs. Moulton, who was my partner at dinner, 1880. seemed to me an extraordinary woman : curiously ugly, but uncommonly agreeable, with something foreign in her manners, with a great range of knowledge—scientific and otherwise—and a great love for it ; has travelled much in various countries —India and Java among others—entirely, it seems, from the love of knowledge. She gave me most glowing and eloquent descriptions of what she had seen in the highlands of Java—riding for a long way together along the edge of the most abrupt precipices imaginable, looking down on the tops of vast forests entirely mantled with brilliant-coloured flowers, amidst which, here and there, masses of Orchids and Ferns, grew as it were, in nests.

June 17th.

We dined with Katharine ; a very pleasant party. The Hookers, Mr. Bentham, Professor Hughes (the Cambridge Professor of Geology), Mr. and Mrs. Sellar, Miss Shirreff; Arthur Lyell did the honours at his mother's table, and did it very well. Bentham, who is just 80 years old, is looking wonderfully well, and is as steadily devoted to botany, and seems as clear-headed as ever. He works six hours every day in the Kew Herbarium. He told me that the last published part of the Genera Plantarum (his and Hookers), concludes the Dicotyledons, and that he is now engaged on the Ochideæ, which are very difficult ; he thinks it probable that they may be the most numerous

1880. family of the Monocotyledons—the most numerous in *genera* if not in *species*. Mr. Hughes (who succeeded Kingsley as president of the Chester Natural History Society), told Fanny that that society is flourishing, and that Kingsley's name is still held in great esteem and veneration at Chester, which I am much rejoiced at.

June 18th.

We had a very large luncheon party—20 including ourselves; — Lady Louisa Legge, the Bishop of Bath and Wells and Lady Arthur, Kate Hoare, Admiral Spencer, Lady Elizabeth Romilly, Lady Muriel Boyle, Sir Lambton and Lady Loraine, Mr. and Mrs. Locker, Mrs. Holland, Miss Frere, Miss Maitland, Lady and Miss Garvock, Minnie, Harry.

The Arthur Herveys were going from our party to hear a lecture from Pere Hyacinthe (M. Loyson); the Bishop told me he had heard him some time ago at Paris, and thought him decidedly the most eloquent man he had ever heard.

June 19th.

A beautiful and warm day, almost the first real summer day since we came to town.

Drove out with Fanny. After a flying visit to dear Mrs. Wilson, and an equally flying look at Waterer's Rhododendrons in Cadogan Place, we drove round Battersea Park (now in great beauty and very enjoyable), and returned by the embankment, skirting Chelsea. I am not sure whether I have ever mentioned in my journal these embankments which are almost (if not quite) the greatest improvements which have been effected in London since I have been familiar with it; improvements both in beauty and convenience, reflecting great honour on the authorities by whom they have been carried out.

June 21st.

A very fine and very warm day. Dear Rose Kingsley arrived to stay with us a few days. I was very glad indeed.

Onr dinner party of eighteen, including Rose Kingsley, the Boyles (Courtenay and Lady Muriel), the Louis Mallets, the Palmer Morewoods, the Lynedoch Gardiners, Lord Talbot de Malahide, and Miss Talbot, Mr. Locker, May Egerton, Minnie, Harry. Very pleasant. I was between two charming women, Lady Muriel Boyle and Rose—a happy position. Patience Morewood was in great beauty.

June 26th.

A violent thunderstorm—the greatest I have witnessed for some years, with an extraordinary downpour of rain which came down in absolute sheets, like a real tropical thunder shower. The thunder began to be heard here about 2 p.m.; all the morning till then there had been a strange, dense yellow fog, like a November fog, growing at last so

1880. thick as to produce almost the darkness of night, so that we had nearly determined on lighting the candles.

———

Dear Lady Grey came yesterday to stay a few days with us. We had a very pleasant small dinner party — Annie Campbell, Emily Egerton, Mrs. Erskine, Lady Grey, Rose Kingsley and ourselves.

———

Dear Rose Kingsley left us. I was very sorry to part with her. Her visit, though too short, has been a great pleasure to me. She is quite as delightful as ever. She gave a very comfortable account of her mother, and spoke quite with enthusiasm of their home that is to be. I only hope it may not disappoint their expectations.

———

Splendid weather. Scott writes to me, " Last week " was very showery, stopping field work to a great " extent : but the temperature continued high, and " *everything* continues to look remarkably well in " the fields. As far as this parish is concerned, I " do not remember ever seeing the corn crops look " better on the whole. I have looked over Mr. " Phillips's, Baldwin's and Cooper's farms, and " consider that the crops look remarkably well. I " am well pleased and perfectly satisfied with the

" prospect on the home farm (and my own farm). 1880.
" At this moment, the wheat and barley are in
" bloom, which is a critical time, and we can only
" hope that the weather may continue favourable
" for another ten days or fortnight, when all danger
" of a deficient yield would be past."

———

June 30th.

Mrs. Adams, a very intelligent and agreeable
American lady, whom we met yesterday at dinner at
Mr. Lecky's, told me that many persons in America,
and even in Massachusetts, are believers in the
divining rod, and (what rather amused me) that the
Witch Hazel of the Americans—the Hamamelis
Virginica—gets the credit in the country of all the
mysterious properties which in Europe are attributed to the true Hazel, though they resemble each
other in nothing but their leaves.

Mrs. Adams seems to have a love of flowers and
trees congenial to mine. She told me that Mr.
Emerson, the author of the beautiful book on the
trees of Massachusetts, is one of her most intimate
friends. She described the delight she had had in
finding the Cornus florida near her home in Massachusetts—for it seems (which I did not know) that
the species is rare in that State. Around Washington (she said) the woods are white with its
blossoms.

———

June 30th.

Went with Fanny and Minnie to the Jermyn

1880. Street Museum of Geology, and showed them the
interesting specimens illustrative of the diamond
deposits and the accompanying formations, as well
as the models of famous diamonds; but I had not
time to find out anything new to me, though there
is always much that is interesting to see in that
beautiful museum.

July 2nd.

Our dinner party, though very large, was very
pleasant; Lord and Lady Aberdare, Lord and
Lady Lilford, Lord and Lady Rayleigh, Lord John
Hervey, Lady Octavia Legge, Sir Francis and Miss
Doyle, Sir Robert and Lady Cunliffe, the Mac-
Murdos, the Clements Markhams, Mr. and Mrs.
Longley. Many more after dinner. Leonora and
her two daughters; also Agnes and Ethel Wilson.

July 4th.

Lord Lilford, when he dined with us on the
2nd, told Fanny that he had that day bought two
eggs of the *great auk*, at Stevens's, for £207. The
species is believed to be now quite extinct.

Mr. Bentham tells me that the frosts of last
winter were very destructive in the Jardin des Plantes
and in particular, that the famous cedar planted by
Bernard de Jussieu, the first cedar planted in
France, has been completely killed.

July 5th.

We went to the Zoological Gardens, but it was a

half-price day, which we had thoughtlessly forgotten, 1880.
and there was such a crowd that we could not see
comfortably. The wonderful *snake-eating* snake,
Ophiophagus Bungarus which I was desirous to see,
appeared lethargic, and showed nothing but its head
and neck from under the blanket; Mr. Bartlett told
us that it had eaten a snake the day before, as
an exhibition for the Siamese ambassador, and this
made it lethargic. Its head and neck show clearly
that it is not of the Viper tribe, but of that group
(of which the Cobra Capello is the most noted
example) which has the external characters of the
harmless Colubridae, but at the same time venomous
fangs. I could not see whether it had the power of
inflating its neck like the Cobra; it showed no
inclination that way. The other serpentine novelty
which I was curious to see, the Echis carinata, would
not show. There are two lively specimens of my
old Cape acquaintance, the Naja, (or Sepedon)
hæmachates; but they are of a nearly uniform dull,
blackish colour, instead of prettily mottled like
others I have seen. A Boa from Madagascar, very
beautifully variegated. An enormous Spider from
Bahia, even larger, I think, than any I saw in
Brazil. The most amusing objects were the
elephants bathing — rolling, and tumbling and
splashing, evidently in great enjoyment.

July 7th.

A sad thing has happened at Barton. We had
the news of it this morning from Scott. A very

1880. respectable tradesman in the village of the name of Knight has committed suicide by throwing himself into a well in our grounds. It appears that he has for some time been liable to fits of nervous depression and morbid melancholy; so no doubt he was not really answerable for his actions. The well is one that I constantly pass in my walks, situated in the grove, which extends to the laundry, and is backed by the holly hedge; there is something rather picturesque in its appearance and situation, and I have been used to call it *Lady Audley's Well*, in allusion to Miss Braddon s novel. I little thought that such a tragedy would be connected with it.

In the terrible explosion of gas which took place two or three evenings ago in the north of London, my nephew Harry had a narrow escape, not as to his life, but as to his property. He had a studio in Fitzroy street, where he carries on his study of art, and has a collection of prints and various interesting objects; the explosion ravaged the street to within 50 yards of that house, and there stopped.

July 8th.

We visited the Royal Academy yesterday afternoon for the second time; went late, from six to seven, and therefore saw it well and comfortably. Much pleased. Besides the pictures I noted on the 14th June, we liked the following :—

"Lady Cunliffe," by *Sant ;* a charming portrait.

"A Thames Backwater," by *Vicat Cole ;* a lovely, peaceful, natural scene.

"A Highland Drove," by *Peter Graham;* one of 1880. his fine pieces of moorland and hill scenery.

"Playful Kittens," by *Faed;* very natural and pretty.

Portraits of Millais and Watts, painted by themselves, for the Uffizii Gallery at Florence.

Some very striking portraits of men, by *Ouless.*

"Watching the Skittle-players," by *Robert Barrett Browning;* a capital picture of a *pig.*

"Sunrise from Waterloo Bridge," by *J. O'Connor* —lovely.

"A Siesta," by *C. Perugini;* a pretty girl in a delicate blue dress of a rather classical shape, reclining asleep in a comfortable corner, with a white kitten on her lap.

"On Silver Thames," by *Vicat Cole;* another delightful scene, quite refreshing to look at.

Portrait of Irving as "Hamlet," by *Long;* very good.

———

July 9th.

A melancholy change of weather since the end of June. On the 1st of this month there was a violent thunderstorm, almost equal to that on the 26th of June, and therewith the weather seemed to break up, and there has since hardly been a day without frequent and heavy showers.

Scott writes that hay-making has been quite stopped. Yesterday about 1 p.m. there was another thunderstorm with torrents of rain.

Lord Lilford, lunching with us yesterday, said

1880. that Yarrell's book is much the best on British Birds ; that Alfred Newton's new edition of it is excellent, but its progress is exceedingly slow, owing to Newton's ultra-scrupulous care and conscientious accuracy.   Lord Lilford spoke with great severity of Morris's work on British birds, declaring that it had done more harm than good to the knowledge of Ornithology.

We visited old Lady Lilford (the Dowager), and had a very pleasant chat with her. She is a charming old lady (by the way, I am not sure that she is really much older than myself, but she appears infirm, and has the manner of an old person). I can remember her very pretty indeed when she was young ; she has particularly kind, gentle, winning manners, and always shows an especially warm feeling towards Fanny and me, on account of my affection for her aunt, Miss Fox, who indeed was loved by all who knew her.

Lady Lilford in her youth saw much of the brilliant, intellectual society of Holland House, and has many interesting reminiscences of the famous men whom she knew there.   I wish she would write them for the benefit of her friends.

July 11th.

Arthur's examination for Sandhurst ended yesterday ; he had been kept hard at work all the week, from Tuesday morning. I hope and believe he will pass, but we shall not know the result for some time. I am satisfied at any rate that he has worked in earnest.

We returned on the 20th from a three days' visit
to the Charles Hoares, at Purley Hall, near Reading
—a place they have hired for the summer.   It was
an exceedingly pleasant visit.   Dear Kate was as
charming as ever—it is impossible to say more—
and both were indefatigably kind and hospitable.
They have three very fine and interesting children,
all boys, the eldest about four and a half; very
pleasing children indeed.   Purley Hall situated
about five miles from Reading, is a rather fine, old
dark red brick house, of the time (I am told) of
James the First, and of the architecture of
that date.   Behind it, facing west, is a steep
grass ascent, variegated with fine trees and crowned
with a beautiful beech wood—very pleasant walks
through the wood and along the brow, commanding
most agreeable views of the valley of the Thames
and the hills on the other side, variegated with
wood, pasture and cultivation.   Just below the hill,
the quiet little village of Sulham, with its spire;
Pangbourne a little further off, nearer to the
Thames ; Mapledurham on the opposite bank of the
river.   In the far distance, a conspicuous blue hill,
crowned with wood, which (we were told) is near
Highclere.

The Hoares gave us two delightful boatings on
the river—up the river to beyond Mapledurham, on
the 19th ; down to Reading on the 20th.   Above
Purley (or rather above Charles Hoare's boathouse,
where we embarked, and which is a good way below
Purley) the scenery is very pretty ; beautiful banks

1880. of wood, and smooth, green slopes, rising in some
places directly from the river side, in others from
the rich, low meadows which border it.   We passed
through Mapledurham lock, and saw   the   quiet
peaceful, little village, and the rustic spire of its
church.   Higher up on the river side, a fine, old
red brick house (Elizabethan or Jacobean) standing
amidst beautiful trees, with a fine wooded hill
behind.   The name (as we were told) is *Hardwick*
(I find it so in the ordnance map, and even in Keith
Johnstone's).

The margins of the river both above and below
Mapledurham, are low, fringed with reeds and
sedges, and with a variety of pretty wild flowers,
which delighted me.   The beautiful blue geranium
I observed only below our starting point, the boat
house.

As we approached nearer to Reading, the scenery
though still pleasing, became less remarkable, the
hills not approaching so near the river.

On the 20th, when our friends took us down in
their boat to Reading, there was a regatta—and
a multitude of boats on the river, and swarms of
people on the bank—a gay scene.   We landed very
near the town, parted with our dear friends and
returned by the railway to town.

The hills here are of chalk, and among the
herbage I noticed several characteristic chalk
plants, in particular, Campanula glomerata (abundant and very fine), Orchis pyramidalus, Origanum
Helianthemum vulgare ; and Asperula cynanchica.

The beech is as I should have expected, the

prevailing tree on these chalk hills, and I saw 1880.
many fine ones. There are also very fine linden
trees now in full bloom and extremely fragrant.

In the beech wood I noticed abundance of wood-
ruff; and I should think that, a little earlier in the
year, one might find there several interesting
plants.

We returned to Eaton Place on the 20th.

The 21st, we attended the wedding (in St.
George's, Hanover Square), of Arthur Wilson to
Miss Kingscote. It was a pretty sight, especially
the bridesmaids; all of Arthur's seven sisters being
in the number. *Felix faustumque sit.*

We all adjourned from the church to an "afternoon tea," at Colonel Kingcote's. July 21st was
our final dinner party for this season—a very
pleasant one.—

Leonora Pertz; Mr. Bentham; Mr. Maskelyne;
the MacMurdos; Mr. and Mrs. Arthur Milman.

July 22nd.—We went with Arthur through part of
the Jermyn Street Museum, and showed him the
beautiful collection of minerals. He was intelli-
gently pleased.

July 23rd.—An extraordinary rain-storm has fallen
at Barton, and done (I fear) a great deal of mischief,
besides the temporary evil of flooding some of the
drains.

Scott writes,—"An extraordinary fall of rain
" occurred here yesterday afternoon (the 21st). It
" came on suddenly in a dead calm from the *north*,
" when the wind was really south-east."

July 23rd.—Fanny and I visited the South

1880. Kensington Museum; saw *Leighton's* new fresco, ("The Arts of War)," which is a fine thing—admired the cartoons and looked at some charming miniatures.

July 24th.—We drove to Veitch's nursery garden, and spent some time there; saw a multitude of curious and beautiful things, especially Nepenthæ, Ferns, and Orchids—ordered several.

Afterwards we visited cousin Emily, who is staying at Norah Aberdare's, and whom we had not seen for several years.

July 25th.—(Sunday). A beautiful day. We had a visit from my dear old tutor, Mr. Matthews, who was very chatty, cordial, and friendly, and (as usual) very quaint. He is in his 82nd year, yet seems to have all his senses and faculties in perfect preservation, and walks alone about the streets of London.

Afterwards, we visited Miss Sulivan, who has a very pretty house and garden on the margin of the Thames at Fulham; indeed so completely on the margin, that when we arrived, the tide being uncommonly high, the river had overflowed more than half of her lawn, and many of her trees were standing in the water. Her house is pretty, and full of works of art; her garden also very pretty; but what struck me most were some exotic trees of remarkable size and beauty; in particular two Liquidambars, the tallest and largest I have ever seen (except perhaps one in the grounds of Sir John Kennaway at Escot, near Ottery); in full perfection of growth and very beautiful; a deciduous

Cypress, taller I think than the one at Embley, 1880.
which I admired in 1878 ; a superb tulip tree ; and
an uncommonly large and fine Ilex.

July 26th.

We returned home to Barton ; had a good
journey, arrived safe, and found all well. Thanks
be to God.

July 30th.

We received, two days ago, the news of the
terrible disaster which has befallen our army in
Afghanistan. There are as yet no particulars, and
there is ground for hope that the destruction may
not be so great as was at first reported ; but I am
afraid it cannot be doubted that General Burrowes's
brigade has been completely defeated and with great
loss. This must, I fear, rouse again all the spirit
of resistance and the animosity against us throughout the whole of that country, and the whole work
will have to be done over again : the whole country
will have to be reconquered,—at what cost, God
only knows. What a host of valuable lives will be
sacrificed ! More than ever, I feel that we ought
never to have gone into the Afghan country.

August 9th.

On the 5th we (Fanny and I), went to Cambridge,
on a visit to Lord and Lady Rayleigh, who are
living in a comfortable small house very near to the

1880. railway station, and we remained with them till the 7th, when we returned home. They were very kind, hospitable and pleasant. In the afternoon of the 5th, Lady Rayleigh took us in her pony carriage to the Fitzwilliam Museum, and we went through the gallery of pictures, which is a large collection, rich in some schools, and containing some interesting things. (There is a good account of this in Waagen's "Works of Art and Artists in England," volume 3, English translation).

I noticed: — "Venus and Cupid," by *Palma Vecchio;* Venus, with a young man playing on the guitar to her—otherwise called the Princess of Eboli and Philip the Second: but Waagen says that the man with the guitar is not at all like the well-known portraits of the King. He says, also, that this is the original of the picture in the Dresden Gallery. "Mercury and Aglauros," by *Paolo Veronese.* A splendid portrait by *Rembrandt,* of an officer in a cuirass, velvet cap and gold chain: to my thinking, one of the finest *Rembrandt's* I have ever seen.

Many excellent Dutch pictures—mostly small.

A large and fine landscape by *Lear*—" The Temple of Bassae in Arcadia." View from the ancient theatre of Taormina in Sicily, by *Linton.* This was interesting to me, because I have so often heard my father say that *that* view from Taormina was the most beautiful scene he had ever beheld.

Afterwards we went with Lady Rayleigh to visit some friends or acquaintances (but found none of them), and finally to see Lord Rayleigh in the great

Cavendish Laboratory where he was at work, 1880. assisted by his sister-in-law, Mrs. Sidgwick. This lady and her husband, together with Mr. Prothero, dined at the Rayleighs—it was a very pleasant party.

On the 6th, in the morning, I walked through the Botanic Garden, but was rather disappointed.

After luncheon, Lady Rayleigh took us to Magdalen, where Mr. Prothero and another gentleman, a fellow of the college, met us, and showed us the Pepsian Library—very curious indeed. We saw one volume of the famous Diary; a very thick, squat octavo in shape, written in a cypher which looks very difficult to make out, but is said not to be so. Also a multitude of fine old engravings, bound in very large volumes. Also a great many curious, old maps and plans and views of London, likewise bound up in large volumes. Also volumes of prints of French costumes; and an astonishing collection of ballads, popular songs and the like.

Of course, though we spent a good deal of time in the library, we could only have a mere taste of its contents. It seems to be well arranged and very carefully kept, and has lately been rendered fireproof. It is certainly a very curious collection, and very significant of the marked and peculiar character of the collector.

This day, the 6th, the Rayleighs had a small dinner party;— Professor Adams (the famous astronomer), Dr. Humphrey (or Humphreys), Professor Clarke, with their respective wives.

The 7th, in the forenoon, we returned home;

1880. Lord and Lady Rayleigh, and Mr. and Mrs. Sidgwick, followed us to Barton in the latter part of the afternoon, and Courtenay Boyle and Lady Muriel also arrived.

————

Mr. and Mrs. Sidgwick went away early yesterday morning; Lord and Lady Rayleigh after breakfast. The MacMurdos and cousin Emily arrived in the afternoon.

Mr. Sidgwick is a remarkably interesting and pleasing, highly cultivated, agreeable man. Mrs. Sidgwick (who is a sister of Lady Rayleigh) very pleasant, and enthusiastically devoted to science.

The accounts from Afghanistan do not seem very comfortable. It is admitted that General Burrows's force was completely, if not disgracefully defeated, and though it is now said that the loss was at first exaggerated, that which is now acknowledged (about 1000 in all, including 21 officers), is bad enough in all conscience. A large force, under Sir F. Ross (a good officer, certainly), has marched from Candahar to retrieve our fortunes there; and we must hope that they will be victorious. But at the same time—what seems very strange—the rest of the army is ordered to evacuate Caubul entirely, and march upon Gundamuck (which is our frontier post). Why the army should make that long march in the hot weather, and why we should make such a decided movement of retreat, just at the time when all the Afghans will inevitably believe it to be the consequence of our defeat on the Helmund,—all this 1880. is hard to understand. If, however, the fact be as Lord Hartington stated in the House of Commons, that these movements have been ordered for strictly military—not political—considerations, and in pursuance of the advice of military authorities, this seems to afford a justification for the Government.

August 12th.

Dear Arthur set out for Scotland, on a visit to Rutherford Clark, a friend of his, in Glenelg, N. of the Caledonian Canal. He (Arthur) has this morning had the news (or rather we have, from Mr. Walrond), of his failure in the second examination for Sandhurst : a great disappointment to us as well as to him. Mr. Walrond, in sending us the news of Arthur's failure, speaks very kindly of him, and expresses confident hope of his success next time.

August 14th.

A very pleasant party in the house for the last week. Courtenay and Lady Muriel Boyle arrived on the 7th, the MacMurdos and cousin Emily on the 9th, Lord Talbot de Malahide and Miss Talbot, Colonel and Mrs. Ives on the 10th. Leonora Pertz and her daughters were here already, having arrived on the 31st July : they and Emily Napier are here now, but all the others are gone. Lady Muriel Boyle is as winning and loveable as ever, and indeed has revealed a charm which we did not

1880. know in her: — her singing of Scottish songs is perfectly exquisite. The MacMurdos are always interesting. By the way, we have lent them our house at Mildenhall for a time, as they have *let theirs* at Fulham. They seem to like Mildenhall very much.

We have been blessed with nearly a week of beautiful weather ; the 9th (except the early morning), the 10th, 11th, 12th, 13th, were delightful days, of brilliant, exhilarating sunshine, and this day, though dull and cloudy, has been free from rain. The weather has been very good for the harvest, which has been going on briskly.

August 19th.

My cousin Emily Napier went away. She is an excellent and very clever woman, warmly affectionate to me and mine, and it is always a pleasure when I have the opportunity of receiving her in my house, which has not happened often of late years.

August 21st.

Another week of fair weather, favourable to the harvest work, which has accordingly made great progress. It is a great blessing. The weather has not, indeed, been all the time what one could call fine or enjoyable ; the 16th, 17th and 18th, were dull, grey, sunless, ungenial days, cold to one's sensations if not cold by the thermometer: but there was no rain nor storm, the weather was favourable enough, and especially convenient to the 1880. harvest men, as it was not too warm for them. Scott gives a very cheerful account of the wheat crop, especially in this parish. Much of it is now carried, and the cutting of the barley is beginning. That crop does not look so well as the wheat, having been much beaten down.

August 29th.

For a wonder we are now actually alone, after having had very pleasant company ever since my note on the 14th. Leonora and her two girls, indeed, have been with us from the last day of July till the day before yesterday, when they set off for Scotland. Besides them we have had, since the 14th, Edward (who went away on the 17th), two Miss Floods (nieces of Lady Cullum), Mr. and Mrs. Sancroft Holmes, Lady Louisa Legge, Lady Florence and Miss Barnardiston, Montagu Mac-Murdo and his daughters Emily and Caroline and Agnes Wilson. Leonora is an excellent and very sensible woman, very well read, very cheerful and pleasant. Her daughters, Annie and Dora, are very intelligent girls, remarkably well educated and well read, with pleasing manners. Annie has a remarkable talent for drawing and painting, and is quite devoted to the fine arts.

Mrs. Holmes, remarkably pretty and very agree-able — altogether fascinating. The two Miss Floods, very clever girls, with lively spirits, active minds, a very considerable amount of knowledge

1880. derived both from books and observation—altogether, interesting as well as entertaining acquaintances.

I had a very agreeable walk with them the day before yesterday, and was delighted with their warm appreciation of Kingsley. On Sunday, the 22nd, I showed part of my museum (principally minerals and fossil plants) to *them* and the two Pertz girls ; the intelligent interest shown by all the four girls, and the appropriateness of their questions and remarks, made it very pleasant to me.

LETTER.

Barton, Bury St. Edmund's,
Sept. 3rd, 1880.

My Dear Katharine,

I did not write to you while Leonora was with us, because I felt sure that she would keep you well informed as to our proceedings. It was a very pleasant time. I was very glad of Leonora's company ; she seemed in very good spirits, and was very agreeable, and so were her girls ; they seemed to be universal favourites, and well deserve to be so. We are just now alone (with Arthur) but shall have a very short time of quiet ; for on Thursday the 9th, the day that Arthur returns to Mr. Hiley's, we go to London on our way to Cheshire, where we are going to visit Lord Tollemache, at Peckforton Castle. It is a sore wrench to leave Barton when it is so beautiful, and we have found it very hard to make up our minds, but on the whole, after long consideration,

we have thought that it may be best both for our 1880.
minds and bodies, to move about a little and give
ourselves some exercise, before the short and dark
days set in and confine us to home. Besides, Lord
and Lady Tollemache have repeatedly invited us in
a very friendly way. I hope you have been enjoying
this glorious harvest weather as much as we have
been, and are enjoying it. It is indeed a very great
blessing to the country, and we cannot be too grateful for it ; and besides its benefits to the country, it
is delightful in the way of enjoyment. The only
drawback is the abundance of wasps, which are an
excessive nuisance.

We went yesterday, four of us, Agnes Wilson, in
addition to ourselves and Arthur, to Mildenhall, and
spent the day there with the MacMurdos very
pleasantly. I never saw Mildenhall look prettier.
MacMurdo took me in his light carriage to " the
hill," where Mr. Betts showed me a Cedar and a
Wellingtonia which quite surprised me : they were
planted by Fanny and me not many years before
we left the place, and I think they are nearly as well
grown as any here of the same age. The common
bracken grows finely in some of these plantations ;
we measured some plants of it eight feet high.

I have not found this beautiful weather with
company in the house, very favourable to reading.
I have only got through part of the third volume of
Spenser Walpole's History. But I go on little by
little, with my *Botanical Fragments* in which I com-
prise what I think worth preserving of my journal
in Brazil.

1880. I am interested in hearing about the progress of the building and the alterations at Kinnordy; I remember that place and all about it vividly, and well I may—and it is with a strange mixture of pleasure and pain that I remember it; so much enjoyment in our first visit; so much sorrow and anxiety in the second. I should like to botanize there again! but that will not be. My love to Rosamond and Arthur, also to Leonard and Mary. I wish you could see Barton now. Everything looks so brilliant in the glorious sunshine, and there is such a variety of beautiful flowers in the long bed where the old greenhouses were; and the new fern-house is succeeding so well this year. I hope we shall not be away more than about three weeks, and that you will come and see us in October.

Ever your loving brother,
CHARLES J. F. BUNBURY.

We have the little Seymours with us. Charlie is a most delightful little fellow.

JOURNAL.

September 6th.

Very good news from Afghanistan. Sir Frederick Roberts has shown himself a really great General, and has followed up a march of extraordinary energy and vigour by completely defeating the enemy in their chosen position; and what is still better, the victory has been gained without any very heavy

loss on our side. I hope this may very soon lead to a 1880.
peace, and to the withdrawal of our forces from that
wretched country. We can now withdraw with
honour. Neither of my nephews were engaged
in the battle; George, who has been seriously ill, is
in the force on its way from Caubul to Peshawer;
William is in Generl Phayre's column, and will be
excessively vexed at being too late for the fighting.

The 9th of September we went to Eaton Place,
and remained there the 10th.

We went by the N.W. railway to Beeston Castle
station, where Lord Tollemache's carriage met us,
and conveyed us to his house, Peckforton Castle;
there we stayed till the 15th. Peckforton Castle is
a modern castle, built by Lord Tollemache, standing
high on the steep, rocky, and wooded promontory in
which the Peckforton hills terminate; looking directly across to Beeston Castle, over the strange
gap which separates the Beeston outlier from the
main range. The castle is built entirely of the pale
red sand-stone on which it stands, and in many of
the rooms the rock is exposed nakedly, without
paper or paint. The great hall and the dining room
are very fine rooms; the drawing room very handsome in a more modern style..

The situation is noble; the hill on which the
castle stands and those which rise behind and above
it, though very steep, are everywhere clothed with
wild and picturesque woods, through which a variety

1880. of delightful drives and walks formed by Lord
Tollemache, lead to many interesting points of view.
The woods are chiefly of oak, with abundance of
fine birches and mountain ash, and a beautiful
undergrowth of heath, furze, bilberry, and fern, very
tall and luxuriant. Bold and massy rocks of red
sand-stone, standing out amidst the woods. From
the prominent points of the hill, there is a very
extensive and magnificent view over the very level
and very fertile plain of Cheshire, to beyond the
Mersey in one direction and to the Welsh mountains in another. I was struck with the remarkable
greenness of the plain—almost all grass land—with
abundance of trees, but scarcely any arable land.

The 14th.—Lord Tollemache took us a long and
tiresome drive of ten miles to luncheon at Dorfold
Hall, Mr. Wilbraham Tollemache's. A fine old red
brick house with the date 1616 on it. A fine
staircase of dark carved oak. Drawing-room (on
first-floor) beautiful and remarkable, especially its
ceiling, which is most richly and elaborately ornamented.

I do not understand, and can find no explanation,
why the Peckforton hills rise so boldly and abruptly
out of the level plain of Chester, being apparently
composed of the same rock, of similar stratification,
and probably of the same age.

Woods around Peckforton Castle occupy about
600 acres; the drives through them extend three
miles.

We were shown, in these woods, a shelf of rock
projecting over a precipice, from which a runaway

horse of Lord Tollemache's leaped down and was 1880.
killed.

The 12th, Sunday.—We went with Lord and
Lady Tollemache to *Bunbury*, and attended divine
service in the church. It is the place from which
our family took its name, but I could not find
any monuments of my ancestors in the church,
which is old and large and handsome. There is
a good monument (a recumbent figure), of a knight,
a Sir Hugh (?) Walter (?) de Calverley, reminding
me of some of those in Salisbury Cathedral.

I had no opportunity of botanizing either at
Peckforton or Bettisfield; but in passing through
the woods around the former place, I casually
observed (besides our two commonest Heaths, of
luxuriant growth, and a Furze, apparently the
common one), also abundance of Golden Rod,
Foxglove, and a large Hawkweed, either Um-
bellatum or Sabaudum. In the roadside hedges
between Peckforton and Bettisfield, Eupatorium
and Angelica in abundance, indicating moisture.
Foxglove near Bettisfield. I was on the look-out
for Clematis, but saw no trace of it. I believe
both places are north of its range.

September 15th.—Lord Hanmer, who had been
staying at Peckforton, took us in his carriage to
Bettisfield, a drive of 15 miles, changing horses
at Whitchurch; the country pleasant, less flat
and monotonous than that on the Chester side of
Peckforton. We stayed there till the 18th, and
then went to Leamington, a disagreeable railway
journey, changing lines at Shrewsbury.

1880.   September 18th to 22nd.—Leamington. We had
come hither expressly in the hope of seeing our
dear Mrs. Kingsley, and we were not disappointed.

On the 19th, we went by appointment and had
above an hour's delightful conversation with her:
and again on the 20th we had a pleasant talk
with her, but it was not so unrestrained and con-
fidential, because her son Grenville was present,
whom we now met for the first time. On the 21st
she was too much fatigued to see us, which was
a great disappointment. These two meetings,
however, were deeply interesting, for we had
neither of us seen her since the death of her
husband. We thought her looking very well,
grown thinner indeed, but not apparently older ;
and in her greeting of us there was all her old
cordiality and warmth of friendship, her old charm
of manner. In her conversation the first day,
though it was delightful, there was a perceptible
tone of sadness, and something of depression, with
perfect self-command ; the second day there was
perhaps more cheerfulness, but still an absence of
that quiet humour and gentle playfulness which
used to be remarkable.

Rose was immersed in the very laborious task of
arranging all their change of home, and preparing
everything for her mother's comfort in the house
into which they were on the point of moving.
She dined with us on the 20th, and was delightful as ever : and the next morning we went to
see her at Tachbrook (their home that is to

be), where she was in the midst of dust and confusion. Tachbrook (which formerly belonged to Walter Savage Landor) is a rather pretty and picturesque many-gabled house, of the time of James the First, not large : in reality, Rose told us, a black and white timber house, but its old exterior concealed under a coat of yellow paint. It has a garden which, I should think, may easily be rendered pretty and pleasant, and some good trees, in particular, a fine Tulip tree. I hope the situation is not damp.

The 20th, we went to see Warwick Castle :— magnificent, both in itself and in its situation. The river Avon flows past it, bordering the pleasure-grounds, between high wooded and rocky banks, the grand and stately towers of the old castle rising most majestically over the abrupt rocks and tufted trees. The view from the bridge is very fine. The approach from the outer gate to the body of the castle, is through a long, narrow and deep cutting in the solid sandstone rock. Magnificent Cedars of Lebanon—some of them as fine (I think) as any I have seen—in the pleasure-ground and close to the castle walls. All the surroundings are fine, and suitable to the majestic building. The view looking down the river, over the green lawns and rich woods from near the conservatory which contains the Warwick Vase, especially beautiful.

(Warwick Castle). The pictures are particularly described in Waagen's "Art and Artists in England," volume 3. I can only note a few here which struck me particularly :—

1880.   "The Earl of Arundel," by *Rubens*—very fine.

"A Dutch Burgomaster," by *Rembrandt*.

"Stafford," by *Vandyck*.

"Charles the First on Horse-back,"—*Vandyck*.

"Robert Rich, Earl of Warwick."—*Vandyck*.

"Ignatius Loyola,"—*Rubens* : I presume a fancy picture, not a real portrait.

Loyola died in 1556.

Rubens was born in 1577.

"Ambrogio Spinola,"—*Rubens*.

September 21st.— We saw Kenilworth Castle ; a fine and stately ruin. It reminded us in some degree of Hurstmonceux, appearing to belong, like that, to the transition stage between the true baronial castle and the mansion. I was struck by the immense size of a mullioned window, and the great height of the walls and towers. The whole is built of red sand-stone, probably quarried in the neighbourhood.

September 22nd.

We went to Oxford to stay with Augustus and Rachel Vernon Harcourt.

October 1st.

From the 22nd to the 27th of September we stayed at Oxford, being the guests of Augustus and Rachel Vernon Harcourt, and we spent a delightful time. I think I noted my great liking for the Vernon Harcourts when they were with us at Barton, in January of this year, and my opinion of them is still higher after spending five days in their

house.   The husband is very clever, as well as
(what clever men are *not* always) very pleasant ; the
wife, clever, accomplished and charming; and both
were indefatigable in doing everything that could
make our stay pleasant and instructive.   Their
house is very agreeably situated on the bank of a
branch of the Cherwell (but not too low) with a full
view of the beautiful tower of Magdalen, and almost
within a step of the bridge, and of the end of the
High Street.   The Augustus Vernon Harcourts
have six children ; five girls, the two eldest twins,
and one boy—very nice little creatures.

I enjoyed Oxford extremely, and saw very much
that was striking and interesting to me : but so
much novelty was crowded into so short a time,
that I am afraid much will have escaped my
memory.

September 23rd.—Fanny went with the Vernon
Harcourts to a party at Blenheim ; I spent the
afternoon very agreeably in the Botanic Garden.
Augustus Harcourt introduced me to Professor
Lawson, who showed me every attention and kindness.   The first and most interesting thing I saw
was the herbarium of Dillenius, containing the very
specimens figured and described in the " Historia
Muscorum."   It occupies much less space than I
should have expected, being contained in one small
cabinet with folding doors.   The plants are fastened
on strong and rather coarse white paper, of small
quarto size (smaller it appeared to me, than the
pages of the " Historia Muscorum,") several species
on one page, but all carefully named and numbered

1880. and arranged in the order of the book.   The names
written in a very clear hand, as if in imitation of
printing.   The specimens I saw are in good (many
in excellent) preservation.   Mr. Lawson told me
that they have the herbarium of Morison, which is
much older, also in good condition.   There are
portraits here of Dillenius (a fat round faced man
—not a clever face) ; of Linnaeus (given by himself)
—Morison and Bobart.   Here also are all Sibthorp's collections ; the whole of Ferdinand Bauer's
beautiful, original drawings for the " Flora Græca,"
part of which Mr. Lawson showed me ; also a great
collection of exquisite drawings (especially of birds,
fishes and serpents) by the same artist, done on the
same expedition, under Sibthorp.   It appears that
there had been an intention of publishing a Fauna
Græca, for which these drawings of Bauer's
furnished ample materials, on the same scale as the
Flora Græca ; but it was abandoned on account of
the expense—and no wonder.

In the garden, it appeared to me, that the houses
were not large, nor of modern construction, and
that the plants in them appeared rather crowded.
But in a large tank in the largest hothouse were
several interesting plants, particularly Pontederia
(Eichornia) azurea in flower very beautiful —and
Salvinia, a very curious, cryptogamous aquatic,
which I had never before seen, either living or
dried—it was here in abundance, covering the surface
of the water in some tubs or boxes, looking at first
sight like a large leaved Duck-weed.   Salvinia of
a paler and more delicate green than in Schkuhr's

plate, and looking flatter and more evenly spread 1880. out on the surface of the water. In the same tank were several of the beautiful blue and crimson Water Lilies of Africa and India ; the Nelumbium with its curious fruits ; the Limnocharis with its delicate pale, yellow flowers ; the curious Water Fern, Ceratopteris, so like the leaves of an Œnanthe.

Out of doors, Professor Lawson showed me the Pilularia (next of kin to the Salvinia) thriving exceedingly on wet bog earth, forming a dense green turf, which might very easily be mistaken for a small growth of grass or rushes—and bearing abundance of its *pill*-like fructification. Also the original Fuchsia on which J. D. Hooker wrote a paper in *The Linnean Society Proceedings*. Also a tree of the *true* Service, Sorbus or Pyrus domestica, derived from the original tree in Worcestershire, This is interesting only for its rarity.

September 24th.

The Vernon Harcourts took us first to the Bodleian. Immensely interesting—the very ideal 1 think of a great university library. There is something wonderfully solemn and impressive in the general effect of these antique galleries and chambers filled with the accumulated learning of so many ages, and in which so many learned and wise men have studied.

Dr. Cox, the librarian to whom the Vernon Harcourts introduced us, was extremely courteous, took us into his sanctum and showed us several very

1880. curious things. Above all, what I thought especially interesting, a number of rough, hastily scribbled notes—mere scraps of paper—which passed between Charles the Second and Clarendon while at Council, and were preserved by the latter. These show a much bolder spirit, are much more blunt and peremptory in style than I should have expected from Clarendon ; they are decidedly honourable to him.

A letter (one might almost say a love-letter), from Charles the First to his wife, in April, 1645 (about two months before Naseby); the hand-writing small, delicate, and beautiful. A MS. prayer-book of Margaret Wriothesley, Countess of Southampton (time of Henry Eighth), with portions of writing in it by several of her friends, as in a modern *album ;* and among these friends were Katharine Parr and Mary Brandon.

Queen Elizabeth's hand-writing (of which we saw several specimens), is very beautiful. There are also a number of beautifully illuminated books. Over the Bodleian Library is a great gallery with many portraits.

Secondly, the New Museum of National History, where Professor Rolleston was our cicerone, and a very agreeable one. I should have been glad of much more time to study the collections, which appear to be very interesting. The room is spacious and well lighted. What seemed to me a very judicious plan is, that the cases with the fossil animals are placed opposite, or near to those containing the nearest recent allies of the same : for example, the fossil crocodiles directly opposite to

the recent.   Here are the typical specimens of the
Ceteosaurus, found in the great Oolite at Enslow
Bridge, 8 miles from Oxford.   This appears to have
been one of the most enormous of all the huge
extinct Saurians.   To give an idea of its proportional
size, a *humerus* of a recent Crocodile of 9 feet long is
here placed in the same case with the corresponding
bone of the extinct monster ; the Crocodile appears
very small and delicate in comparison.

The Ceteosaurus is very fully and elaborately
described and illustrated in Phillips's "Geology of
Oxford," pp. 245-294.

The Dodo ; the head and foot preserved here are
some of the greatest curiosities, being all that
remains of the original specimen formerly preserved
in the Ashmolean Museum.   In the same case a
skeleton of the Didunculus, which appears like a
miniature Dodo.

Teleosaurus (Gavial)—several fine fossil remains
from the neighbourhood of Oxford.   Professor
Rolleston remarked to me, that the Gavials, in the
Jurassic age, were evidently abundant and widely
distributed : whereas now they are confined to the
Ganges, the Indus and a few other Indian rivers, so
that they appear to be a form which is gradually
dying out.

Specimens of jaws of the Stonesfield Mammal,
which was long supposed to be the most ancient of
Mammals.

Professor Rolleston pointed out to us a very fine
skeleton of a Python, showing the curious little
bones which are the rudiments of legs.

1880.   Thirdly, New College. In spite of its name, this was founded in 1380, by the famous William of Wykeham: and according to Parker's Handbook for Oxford, the buildings remain for the most part as they were erected in the founder's time and on the founder's plan. We admired the beautiful chapel and anti-chapel, with their rich coloured glass windows, portions of which are of the time of the founder.

The great west window was painted by Jervais, from finished cartoons furnished by Sir Joshua Reynolds.—*Handbook for Oxford.*

According to the Handbook, the "ribbed" roof of the cloister "resembling the bottom of a boat," is of Spanish chesnut wood. The beautifully wrought altar-screen (restored by Wyatt),—the cloister, with the curious ribbed roof, the fine and lofty hall, and the beautiful garden, partly enclosed by the old battlemented city wall, which has a walk along the top within the shelter of the parapets. In this and some other colleges, particularly Magdalen, we were struck by the beautiful effect produced by the creepers (Virginian and Japanese), planted against the old walls, and mantling them with rich masses of foliage, now of all shades of crimson and green.

The tower of New College is "supposed to have been the last work of Wykeham. It is built on the site of one of the bastions of the city wall, and as its massive nature evidently imports, was intended for defence as well as for a belfry."—*Handbook.*

Fourthly, we saw and admired All Souls. We

had already admired its long and stately front on 1880.
the High Street. The chapel is remarkable for its
extremely beautiful and elaborate reredos, which
has been restored within the last few years by Sir
Gilbert Scott, from some portion which remained of
the original structure. The library, a grand room or
gallery, of great height and prodigious length ; here
are preserved a great quantity of original plans,
drawings and studies by Sir Christopher Wren, who
was a member of this college. All Souls was
founded in 1437, by Chicheley, Archbishop of
Canterbury. (In the Handbook).

This interesting day ended with a pleasant little
dinner party at the Vernon Harcourts: Mr. and
Mrs. Max Muller, Mrs. Pattison, Mr. and Mrs.
Child, Professor Lawson.—All agreeable, especially
the Max Mullers. I had long wished to know *them*,
and I was not disappointed. Mrs. Muller is very
handsome and very agreeable.

September 25th.

We saw first, Christ Church—magnificent indeed.
The great quadrangle, measuring 264 feet by 261, is
one of the finest things of the kind I have ever
seen ; I hardly think that even the Great Court of
Trinity at Cambridge comes up to it. The tower
surmounted by a cupola, and containing a great
bell, is a striking object. The hall is reached from
the great quadrangle by a grand staircase ; the roof
of the lobby which contains this staircase is
extremely beautiful, of the finest fan tracery,

1880. radiating from a single supporting pillar. The hall, a most noble one, almost superior to that of Trinity College, Cambridge. All along the walls, a series of interesting portraits of eminent men who have been connected with the College. We remarked particularly: — Lord Wellesley, Lord Mansfield, Lord Grenville, Bishop Trelawny, Archbishop Longley, George Grenville (the Prime Minister of whom Macaulay draws such an unfavourable picture), Dean (Cyril) Jackson, Lord Auckland (the first), Canning Busby (the notorious schoolmaster), Bishop Goodenough (the botanist), John Locke.

The library of Christ Church (which, as it happened, we did not see till the 26th), is grand, and appears to contain a noble collection of books. There is something very impressive and interesting to me in these great college libraries, brought together and preserved by so many generations of studious men. On the ground floor beneath the principal library is a considerable collection of pictures, given to the college by General Guise; there is a catalogue of them in the Handbook for Oxford.

The second quadrangle of Christ Church, called Peckwater, is very large and handsome, though not equal to the first. There is a third quadrangle, much smaller, named Canterbury, from its occupying the site of a former Canterbury College. It happened somehow that I did not see the interior of the cathedral, which is also the chapel of Christ Church. I was very much impressed with the

beauty and solemnity of the Christ Church walks—
those magnificent avenues of elms.

Next, we visited Magdalen, which is one of the most beautiful colleges; especially remarkable for its magnificent tower, which in all points of view (especially in combination with the bridge), is one of the finest objects in Oxford; secondly for the beautiful cloisters; thirdly for the strange assemblage of stone figures of various extraordinary and grotesque monsters, ranged round the interior of the quadrangle of the cloisters*; fourthly for its beautiful "water walks," bordered by branches of the river Cherwell, shaded by stately and venerable elms, and affording beautiful views of the college from some points, and of the bridge from others. One part of this walk is called Addison's Walk. The "grove, or deer park," is also remarkable, from its rural appearance and its many fine old trees.

Merton College.—Library very old, quaint and curious; said to be certainly "one of the oldest, and perhaps now of the most genuine ancient library in this kingdom; it has a good boarded ceiling of the 15th century, divided by mouldings into small square panels, with bosses at the intersections, painted with small shields of arms."—*Handbook*.

The garden interesting,—enclosed by a portion of the old city wall, "forming the south-east angle of the fortification." An agreeable walk along, or

---

* It is doubtful whether the monsters have allegorical meanings or are mere grotesques. The cloisters have been in great part restored within this century.—*Handbook*.

+ Magdalen College was founded by Bishop Waynflete in 1456; built 1475-1481; the tower added 1492-1585.

1880. just within the top of the wall, shaded by fine trees and affording pleasing and interesting views.

Merton College.—Said to be (at least as to its foundation), the most ancient of the colleges ("founded at Maldon by Walter de Merton, 1264—transferred to Oxford, 1274."—*Handbook*).

Chapel very beautiful.

After Merton we saw Wadham College. This has a very handsome front, but we were chiefly impressed by the beauty of its garden : not large, but delightfully quiet, verdant and fresh, with several trees of remarkable beauty and most happily placed. The lime trees, cedars, and purple beech are magnificent.

———— —

September 26th.

Fanny went with the Vernon Harcourts to divine service in the Cathedral. I did not stay there for fear of cold. In the afternoon our friends took us to Christ Church, which we had only partially seen before ; then to Pembroke, Worcester, and Exeter Colleges, ending with a look into the quadrangle of St. Alban's Hall, to see a Wisteria of extraordinary size.

Pembroke interesting as Samuel Johnson's college —a fine portrait of him (by *Reynolds*) in the Master's lodge.

Other famous men who were educated at Pembroke College were Sir Thomas Browne (" Vulgar Errors"), Judge Blackstone, Whitfield, and Shenstone.—*Handbook*.

Worcester College, founded by Sir Thomas

Cookes in 1714.—"Although the most modern but two of the existing colleges, it occupies the site, and in its buildings exhibits the remains, of one of the earliest foundations for religious learning in Oxford."
—*Handbook.*

The original foundation was in 1283, by John Giffard, Baron of Brimesfield. It was called Gloucester Hall.

Worcester College, remarkable for the beauty of its garden, rich in fine exotic trees, and bounded on one side by a pretty sheet of water, which I was told, afforded in winter the only good skating at Oxford. The Chapel, quite modern, but remarkably rich and beautiful, the walls painted throughout, and, together with the coloured windows, the roof and floor and all the decorations, combining to illustrate the one idea, of all creation praising the Lord. The paintings are very beautiful. The only objection I perceived was, that the intensely coloured windows admit so little light, that unless the door be open, one can see hardly anything.

The designs of this Chapel are fully explained in Parker's Handbook for Oxford.

Besides these, there are other interesting colleges and academic buildings, of which (for want of time), we saw only the exterior: in particular, Balliol, Oriel, Corpus Christi; St. Mary's (the Church of the University), of which the magnificent spire is one of the most conspicuous and commanding objects in all views of Oxford: and the Radcliffe Library, the cupola of which is another most prominent and striking object.

1880.   The High Street of Oxford appeared to me fully
deserving of its fame; I certainly do not remember
any other street view in England comparable to it.
Its width, the gentle inclination, the graceful curve
in its direction, the noble buildings which succeed
one another almost without interruption as one
passes along it, the absence of meaner features and
its beautiful termination in the tower and bridge of
Magdalen, excel anything else of the kind that I
remember.   Certainly Cambridge will not bear a
comparison.

St. Mary's:—Tower and spire built about 1300,
chancel 1460, nave about 1488. — *Parker's Handbook.*

The panels and gables of the pinacles testify to
its date, being lined with a profusion of pomegranates, in honour of Eleanor of Castille, mother of
Edward II., in whose reign it appears probable the
work was complete.—(ib.)

The only thing I saw objectionable in Oxford was
the new suburb,—ugly, mean and disagreeable,
between the railway station and the upper end of
the High Street.

Oxford, like Cambridge, lies rather low, but has a
great advantage in its rivers, being almost enclosed
as well as penetrated by several branches of the
Thames (or Isis) and the Cherwell, which are very
superior indeed to the ditch-like Cam.

———

September 27th.

We went from Oxford to London; and on the
29th came home to Barton.

While we were at Oxford we drank tea with Mr. 1880. and Mrs. Max Muller, and found them both extremely agreeable.

======

LETTER.*

My Dear Katharine,

Many thanks for your kind letter of the 20th, which I had intended to answer sooner, but first some very disagreeable private business, and since, out-door occupations and business with tenants here, have caused this delay. We came hither on the 23rd, together with Joanna and Dora, Arthur and his tutor Mr. Trevor. We have been very busy visiting some tenants and receiving visits from others, inspecting the plantations, new churches, and school-houses, and the like.

The young plantations are flourishing, which cannot be said of everything here ; the wet season, which has disagreed so much with all other kinds of crops, has agreed very well with them.

Altogether, though I have very many agreeable and interesting recollections associated with this place, this visit has been one much more of duty than of pleasure. It is indeed an unlucky season for visiting one's tenants. I am afraid Joanna and Dora must have found Mildenhall very dull.

I have a very curious Fern to show you when you come to Barton ; very possibly indeed, you may

* This letter should have appeared in p. 159 of this volume.

1880. have it among your Bornean specimens, but it seems to be very rare; it was sent to me by Dudley Hervey, who found it on a high mountain in the territory of Johore, which is the part of the Malay Peninsula nearest to Singapore. It is Acrostichum *bicuspe* of the species Filicum, and *Gymnopteris Vespertilio* of Hooker's *London Journal Botanical*, where there is a very good print of it.

I lately bought (from Quaritch) a large collection of Sicilian dried plants, the arrangement of which will furnish me with entertainment probably for the rest of my days. They are very good specimens and all named by Professor Todaro. I delight in the plants of the Mediterranean region.

Frank and his bride paid us a very short visit, but my impression of your new daughter was decidedly pleasant, and I have no doubt that I shall like her much when I have the opportunity of being better acquainted with her.

We had a pleasant visit at Helmingham, Lord Tollemache's, a very interesting old place; but you will have had a full account of it in Fanny's journal and as I hope to see you soon, I will write no more now. The harvest has been deplorable, especially in the fens, and I have no doubt there will be a great falling off in my rents.

Ever your loving brother,

CHARLES J. F. BUNBURY.

October 2nd.

Barton,
October 5th, 1880.

My Dear Joanna,

I thank you very much for your very pleasant and interesting letter of the 12th of September, from Villars. I suppose you are all by this time re-united at Florence, and I hope none of you the worse for your summer ramble.

I have not seen the botanical book (Kitchener's) which you mention, but will look out for it, as I often wish to be able to recommend some clear and simple and at the same time *sound* elementary book on the subject. Joseph Hooker's is very difficult, at least, parts of it. I will set about making a selection of English plants from my duplicates for you, and hope I may have a good *lot* ready against the next opportunity of sending them to you. But I do not know what help you may have for Italian botany, which is incomparably richer than ours. Is there any good elementary book in Italian, on botany?

I very much agree with you as to Justin M'Carthy; his two volumes are very pleasant reading, a particularly pleasant style—and I, like you, was struck by the fairness and candour with which he generally writes on persons and events, a quality the more remarkable as he is, in Parliament, a determined *Home Ruler*, and (I am told) a very troublesome and impracticable one. I quite agree with you that M'Carthy has not done justice to Prince Albert and the Exhibition; and I dare say you may be right as to his not feeling sufficiently the evil

1880. influence of Popery: but in England this is not brought so much under our notice.

I am just finishing another work on nearly the same period as Justin M'Carthy treats—namely, Spencer Walpole's History of England from the close of the great war. This also is interesting, and carefully and conscientiously done, I think; but it is hardly so attractive in style as the other, and here and there, especially in the 3rd volume, are passages in a tone I do not quite like, rather bitter and censorious. He is also too much of a Radical for me, though his father is a Tory. But the book is very useful. The only novel (except W. Scott's and a few other old ones) which I have read for a long time, is *Le Blocus* one of the Erckmann Chatrians set, which interested me very much; but I have not room or time to discuss it now. I hope you are all well at Florence, as we are I am thankful to say. With much love to all your party.

Ever your loving brother,
CHARLES J. F. BUNBURY.

Oct. 7.

Barton, Bury St. Edmund's,
October 25th, 1880.

My dear Katharine,

Do you remember giving me (last year, I think), some seeds of a leguminous plant, which had been sent to you from Natal, but were said to have come originally from Madagascar? They have grown and flourished, and one of the plants

has now blossomed, and is very curious and 1880.
beautiful; the flower papilionaceous, in colour lilac
and white, all the petals wreathed in a spiral, in a
most singular manner, the bud just before ex-
pansion, looking exactly like a snail shell. It is the
Phaseolus Caracalla of Linnæus, formerly known in
English gardens as the Snail-flower, said to be a
native of India. I do not know whether it is
annual : if it is I am afraid we shall not keep it
long.

Fanny's journals gave you such a good and full
account of our little tour, that I need not say much
more about it.

I was delighted with Oxford, and enjoyed
thoroughly our 4 or 5 days there, but all that I can
tell you in addition to Fanny's description of the
Colleges, &c., is about the Botanic Garden. It is
not a large one, nor very rich,—looks (like that at
Cambridge), as if it would he the better for the
expenditure of some more money upon it ; but it
contains many interesting things. The Professor,
Mr. Lawson, seems clever and very zealous, and
was extremely courteous and attentive and pleasant
to me. He showed me part of the famous
Dillenius's herbarium of Cryptogamous plants,
containing all the original specimens figured in his
Historia Muscorum, most neatly arranged and
labelled, and in very good preservation, though
gathered at least 140 years ago. I saw also some of
Ferdinand Bauer's exquisite drawings for the Flora
Græca ; the whole are there, as well as a great
collection of equally elaborate drawings, by the

1880. same artist of the animals of Greece and Turkey ; the fishes and reptiles most beautifully done.

Sibthorp's intention had been that a Fauna Græca should be published from these materials, on the same scale as the Flora : but the funds were not sufficient.

Of living plants, what interested me most was the *Salvinia natans*, flourishing in abundance in a tank in the hothouse ; a very curious floating Cryptogam, which I knew before only by descriptions and plates : there are illustrations of its structure in Maout's and Decaisne's book (which you know in Mrs. Hooker's translation), but the figure gives no idea of its general appearance. Another curious plant which Mr. Lawson showed me was the Pilularia, which, though it is an English plant I had never before seen alive.

In the New Natural History Museum, Professor Rolleston (a very pleasant man), showed us several interesting things, but I was sorry not to have more time to spend there. The greatest rarity there perhaps are the head and foot of the last of the Dodos—"the much lamented Dodo," as Hallam calls it. It is long since I have spent four days in a way so full of interest and instruction, and with such an acquisition of new ideas as those at Oxford : with the great additional advantage that we were staying with delightful people.

Since we came home we have had a constant succession of pleasant guests : but I refer you to Fanny's letters for information about them.

I hope that you and all your children and grand- children are well.   I was very glad to hear of Frank's happiness.

Ever your loving brother,
CHARLES J. F. BUNBURY.

JOURNAL.

November 1st.

Almost through the whole of last month we have had a continual succession of agreeable company in the house.   Taking them in the order of time :— Mrs. Ellice, and Helen, the Louis Mallets, John Herbert, Admiral Spencer, the Edward Goodlakes, Mr. and Mrs. William Hoare, Kate and Charles Hoare, William Napier, the Boyles, (Lady Muriel and Courtenay)— and lastly our Bury ball party, consisting (beside the Boyles), of the Morewoods, Caroline Hervey, Lady Evelyn Finch-Hatton, Sir John Shelley and his sister, Cissy and Emmy, Charlie Napier, Lord John Hervey, Miss Tolle- mache, Mimi MacMurdo, and Mr. Denman.

Dear Minnie Napier has been with us the whole time and is so still.

This last party was an especially gay and lively one, with a remarkable share of beauty in the ladies.   Most of them (I mean of the whole company from the beginning) are old friends, whose merit I have often recorded.   I need not expatiate in praise of Kate Hoare, the Boyles,

1880. the Louis Mallets, Mrs. Ellice, Lady Evelyn, the Shelleys.

October has not been this year, a month of fine weather, as it often is here. There was a great deal of rain in the early part. Several fine days in the middle time, but on the 20th and 21st there was actual snow.

The autumn colouring of the trees here is always beautiful, especially about the pleasure-ground and arboretum and the groves near the house, where my father had particularly studied this beauty in designing his plantations : but the colouring has certainly not been so fine as usual. The Virginian creeper and its Japanese congener, which we saw in such beauty at Oxford, have not been so richly coloured here. The Beeches, however, both the common and the purple, are very beautiful. The American Red Maple, which delighted us all by its brilliancy last autumn, is hardly less vivid in colouring this year.

The rainfall on October 3rd (*i. e.* in the 24 hours ending on the morning of the 3rd), was 41 inch;

The 5th	....	....	·44 inch.
The 6th	....	....	·22 ,,
The 7th	....	....	1·00 ,,
The 9th	....	....	·14 ,,
The 21st	....	....	·24 ,,

This being partly snow.

The 27th	....	....	·80 ,,
The 28th	....	....	·23 ,,
The 29th	....	....	·20 ,,

If I am not mistaken this is considerably above the average of our October rainfall. The month, however ended, and the present one has begun, with very fine and bright though very cold weather.

On the 21st of October, there was a public meeting at Bury, on behalf of the Suffolk branch of the Society for Prevention of Cruelty to Animals. The Bishop of Ely was in the chair, and gave us a really beautiful and impressive address; his language, voice, and manner, all so winning, that it was a real pleasure to hear him. Several other gentlemen spoke well. The meeting was in every way a success, and will, I hope, do good. Fanny had been indefatigable in gaining recruits for the cause, both by letter and personal application.

November 4th.

Received a letter from Edward, from Athens —this was a great relief and comfort, for we had not heard from or of him for nearly a month, and we had expected to hear much earlier. His letter gives a very satisfactory account of himself, and is altogether very agreeable. He had made (by steamer) the circuit of the Ægean, from Constantinople by Salonica (Thessalonica) to Athens: had enjoyed beautiful weather all the time, and had had a perfect view of Mount Athos. He says—" One of " the principal objects of my present tour was to " make the circuit of the Ægean Sea, from the " Dardanelles by Salonica to Athens; and I made it

1880. " under the most favourable circumstances possible,
" not only having perfectly fine weather, but an
" accidental delay having prevented our passing
" Mount Athos, as the steamers usually do, in the
" night ; instead of which, we passed it in broad
" daylight, and on account of the perfect calmness
" of the sea were able to steam close round it.   It is
" certainly the finest mountain promontory in
" Europe; and the multitude of small monasteries
" hanging on its sides, and perched in the most
" inaccessible situations, gives it a very curious
" appearance, even as seen from the sea."

Of Athens, from whence he writes, he says that
he has been much interested, not only with seeing
the ruins and ancient remains again, " and a good
" many monuments which have been discovered
" since I was here last, but also in seeing the
" changes and improvements that have taken
" place since then, and which are truly astonishing.
" Whole streets of very good houses have grown up,
" and many of these which have been built by
" wealthy Greek merchants and bankers, are really
" splendid.   At the same time many of the poorer
" quarters, which used to be wretchedly bad, have
" been either swept away, or much improved ; and
" although some portions remain to remind one of
" what Athens *was*, the greater part of it is now a
" well-built European-looking city.   The Piræus
" also, where, after the close of the War of Indepen-
" dence, there was not a single house standing, is
" now a thriving, bustling, commercial town, with
" factories and magazines rising on all sides : not a

" pleasant place to reside at, but still very different 1880.
" from Pera or Galata."

From the 30th of last month (inclusive) to
yesterday, continued fine weather, indeed beautiful
—clear air, almost cloudless blue sky, and brilliant
sunshine; wind (when one was exposed to it) very
keen, N. or N.E.; frost every night; thermometer
down to 22 deg. in the night of the 1st to the
2nd.

These frosts have brought down the leaves in
heaps, and killed off the last lingering flowers in the
garden. The beeches have still a great deal of
beauty, and some of the oaks and elms have hardly
even yet changed colour; but otherwise almost all
the deciduous trees are leafless.

I had written to Kinglake on the 10th, to tell him
my opinion of his new volume, and this day I
received a very pleasant letter from him, expressing
himself much pleased with my approbation.

Saw in the newspapers the death of the Lord
Chief Justice Sir Alexander Cockburn, at the age
of 78. I can hardly say I was acquainted with
him, except that I repeatedly dined with him,

1880. when he gave his official dinners on occasion of the Assizes at Bury.   He was the principal Judge at Ipswich, at the summer Assizes of 1868, when I was High Sheriff of Suffolk; and I was exceedingly impressed by the ability and acuteness of his summing up on one of the principal trials (a charge of rape)) ; it was masterly.   I have understood that his moral character was by no means as high as his intellectual.

———

We received the news in a letter to Fanny from Mrs. Godfrey Lushington, of the death of our dear old friend, Mr. Samuel Smith, at Embley near Romsey.   He has lived to a great age—86 years, I believe.   Mr. Smith, according to the newspaper was in his 86th year.   He was a son of William Smith of Norwich, a famous Whig of the last generation.   He married the sister of Mr. Nightingale, of Embly in Hants and Lea Hurst in Derbyshire;   and Mrs. Nightingale was Mr. Smith's sister.   He and his family were old and constant friends of the Horners, and he always showed a cordial feeling towards Fanny, and to me for her sake.

Mr. Smith has for some years been very infirm, so that one ought not to sorrow for his death ; but I remember with a melancholy pleasure, the delightful hours which I have formerly passed in his company.

We visited him and Mrs. Smith at Embley, two

years ago, in August, 1878, which was the only time I have ever seen that beautiful place; but he was then very infirm, though his mind was perfectly clear. I gladly associate him rather with recollections of his former home, Combe Hurst, near Wimbledon, where we visited him often, and enjoyed at once that lovely spot and his delightful society. Delightful indeed he was in those days. I have seldom known any man more entirely to my taste. With a highly cultivated mind, and abundant knowledge both of men and books, he had a peculiar, unaffected suavity, a mildness relieved by a playful, delicate humour, which gave a remarkable charm to his conversation.

Combe Hurst was just on the verge of Combe Wood; rather indeed, a bit out of Combe Wood, enclosed and cultivated; it stood high on the brow of the hill, looking in one direction across to Wimbledon Common; while on another side, the view ranged down a beautiful sloping lawn, fringed with flowering shrubs and flanked by woods, into a valley, and beyond to a succession of hill and dale extending far into the distance. The evergreens were luxuriant, the rhododendrons and azaleas most beautiful (for the soil exactly suited that family of plants), and from the flowery lawn one passed at once into a charming mixture of heath and wood, with fern and foxgloves flourishing under the oak trees. It was indeed a delicious spot.

December 2nd.

A disagreeable thing has happened next door—as it were—to us. Nether Hall, Lady Hoste and Mr. Green's house, at Pakenham, was entered by burglars in the night of the 29th of last month, and the whole of the plate—which was of great value—carried off, as well as bonds, to the amount of £3,000.

December 6th.

Our dear Arthur returned from London, his examination being over : but it will be a good while before we shall know the result.

December 16th.

Dear Harry left us, to go at first to London and then to Antwerp, where he intends to settle himself for some time, to study painting in earnest, as he has determined to devote himself to the profession of an artist. He is an excellent fellow, and I heartily hope he will be successful in his object.

The Luculia gratissima is now in blossom in the conservatory (brought thither from the hot-house), very beautiful and fragrant.

The conservatory is very gay with flowers, particularly with the scarlet Poinsettia and Salvia splendens : the Salvia particularly ornamental, because its flower stalks and calyxes are of as brilliant a colour as the corollas, and therefore its splendour lasts much longer than that of most flowers.

A very cold day — a strong white frost in the morning. I began throwing out bread before my windows for the birds : and among the very first which came, was a blackbird with a white patch on each cheek—to all appearance the same identical bird which has visited us each of the last two winters.

Sorry to see in the papers the death of Frank Buckland. I did not know him personally, but he is a man to be regretted :—a good naturalist, a pleasant writer, and very useful by his knowledge of everything relating to fish, and his practical application of that knowledge. What he has done in promoting *fish-culture* is, I believe, very valuable.

We had the news, a few days ago, of the sudden death (at Florence) of Lord Crawford and Balcarres, better known (to us at least) as Lord Lindsay. His " Lives of the Lindsays " is an excellent book : his " Christian Art " very interesting.

Deplorable news from South Africa. A detachment of the 94th regiment, 250 strong, has been overpowered (surprised, I suppose), by the insurgent Boers of the Transvaal :—one officer killed, the

1880. commanding officer, Colonel Anstruther and 3 others badly wounded, 120 men killed or wounded, and the rest of the party taken prisoners. It is very painful, and what makes it more so, is that (as it appears to me) we are in the wrong, and are engaged in an unjust war with these Transvaal Boers. But now, as blood has unfortunately been drawn, and our troops have met with a check, we shall, in a manner be obliged to go on till we have crushed these poor, honest, brave men, or driven them further into the interior.

———

December 29th.

It is now said that the first accounts of the disaster in South Africa were exaggerated,—that only 35 men of the 94th were killed or wounded, and that those who had been taken prisoners, were released and allowed to proceed to Pretoria. It is, however, also asserted, that the attack was an act of treachery on the part of the Boers, for that the time had not yet elapsed which it had been agreed, should be allowed for the answer to their demands. One can hardly tell what to think till we have more complete and trustworthy information.

———

December 31st.

The greater part of this month has been damp and tolerably mild, but it is ending in true wintry fashion.

We are now very near the end of another year, and again I feel myself bound to offer up most humble and heartfelt thanks to the Almighty for the many and great blessings which He has granted to me. Above all, I can never be sufficiently thankful that my invaluable wife is preserved to me, and that we live together in unbroken peace and harmony.. In spite of many sad losses, we may also well be thankful for the affection of so many kind and valuable friends, as well young as old. As to my health, it has not been as perfect as it was a few years ago, but I have still, considering my age, great reason to be thankful. I have not expressed half strongly enough the gratitude that I feel, especially for the inestimable blessing of unimpaired eyesight, and for the preservation of my mental faculties, the power of enjoying books and nature alike.

Dr. Andrew Clark—when he examined me just before we left London—pronounced me to be in all essentials healthy and sound. I am still more grateful for the good health of my beloved wife.

We have lost, in the course of this year, some old and valued friends:—Mr. Samuel Smith, Lord Charles Hervey, and our excellent old servant Mrs. Marr.

Mr. Smith indeed was so very old, that his death was reasonably to be expected, and could hardly be otherwise than welcome to himself. Lord Charles I regret very much. Sir Henry Blake, though not very congenial, was a friend of very old standing, a good man and always friendly in his sentiments and conduct towards me. Mrs. Abraham also must be reckoned among the friends we have lost.

1880. Lady Augusta Seymour we knew less well, but as far as our acqauintance extended, I had a great liking for her, even independently of the warm affection between her and her brother, our dear Lord Arthur. Lady Hanmer also must be added to the list.

1881.

JOURNAL.

January 1st.

1881. Two very agreeable episodes in our personal history in the past year were our visit to the Charles Hoares at Purley Hall, in July, and that to the Augustus Vernon-Harcourts at Oxford, in September. I shall remember both with pleasure as long as I live. The former, in addition to the enjoyment of dear Kate's delightful society, made me acquainted with beautiful and interesting scenery which was new to me ; the other enriched my mind with a multitude of new and interesting images and thoughts, which it will always be a pleasure and an advantage to recal. It is long indeed since I have had such acquisitions to "the furniture of my mind," as in that visit to Oxford.

I must not however forget our trip to Cambridge when we were the guests of Lord and Lady Rayleigh (which also was very interesting).

The Oxford expedition gave us also another advantage in bringing us acquainted with Professor and Mrs. Max Müller, whom I had long wished

to know, and whom we found charming ; and with 1881.
Professors Rolleston and Lawson.

In respect to public affairs, the past year has been
by no means a bright or cheerful one. Though the
weather was very fine during the time of harvest,
the crops (probably owing to a bad season for the
blossoming of the wheat and barley), proved by
no means equal to what had been expected. At the
same time, the enormous importations from America
caused the prices to be very low ; so that the unfortunate farmers have been under the disadvantages
at once of a deficient crop and of low prices. The
prospect is indeed very dreary, both for farmers and
their landlords. More favourable seasons may
come, and probably will, but the high prices most
likely will never recur ; nor ought we to wish that
they should, for cheap bread is a primary requisite
for the hard-working classes.

As to politics, the whole state of affairs is to
my thinking so detestable, and the prospects so
dismal, that I will not dwell upon them. In the
place of one war in South Africa, as last year,
we are now engaged in two at once, and each seems
likely to require the earnest exertion of our strength;
and, what is the worst, it appears to me that we are
in the wrong in both. At the same time, the war
in Afghanistan, though for the present dormant
or smouldering, is likely enough to break out again
as soon as we withdraw our army. But by far
the worst part of our situation is the condition of
Ireland, whether considered in itself or in its
probable effects upon England. In either view, it is

1881. disastrous and disgraceful. Such a sanguinary and dreadful anarchy is worthy only of Mexico; or of France in the worst part of its first revolution. I well remember how, when I first began to understand and to take an interest in politics, the Irish question was the great difficulty of politicians; between fifty and sixty years have passed, Irish questions have been incessantly the difficulty and plague of politicians and the state of Ireland really seems to be worse than ever. It appears to be almost absolutely hopeless. Very lately indeed there have been some signs which seem to show that the Ministers are sensible of the shocking state of Ireland, and wish to do something to remedy it; but I am afraid it is too late. I doubt whether even Parnell and his gang could now restore law and order.

The look-out is very dark indeed; we can only pray that God may help our country and do better for us than we can do for ourselves.

———

January 6th.

Fanny has made me a splendid present, of Elwes's magnificent book on the "Genus Lilium," the most beautifully illustrated botanical work I have seen for a long time.

We have very good accounts of Harry, from Antwerp. He has got comfortable lodgings, has some acquaintances whom he likes, has entered the Academy, and works there regularly for several

hours each day. I trust he is now fairly launched, 1881.
and will persevere and succeed.

January 10th.

We had the good news that Arthur has passed
his examination for Sandhurst. Mr. Walrond very
kindly telegraphed the news to us, knowing Fanny's
great anxiety on the subject.

January 14th.

The young ones of our party here, yesterday
evening, acted a charade very well: it was very
well got up, and very amusing. The performers
were :—Rosamond Lyell, two of the Wilson girls,
Dorothy Hoste, Arthur, the three young Walronds,
and Hedley Strutt.

The Word *Misselto*, divided thus—Missile—Toe.
The *first* scene was the Carnival at Rome, and the
pelting with comfits *(missiles)*: the *second* was taken
from the fairy tale of Rumpelstiltzkin; the *third*
exhibited the wedding of the unfortunate bride who
perished in the "old oak chest," according to the
legend, and the song of "The Misselto Bough" was
sung by the performers. This scene was very
prettily and cleverly got up, and the bride was well
represented by Dorothy Hoste.

January 15th.

The pleasant party who have been with us for
these five days, broke up to-day. They were :—
Katharine and Rosamond, Mr. and Mrs. Walrond,

1881. and *his* three sons by his first wife, Mr. and Mrs. Hutchings, Lady Rayleigh and her youngest son Hedley, Ethel and Constance Wilson, Dorothy Hoste, and my nephew Cecil. The young people were very gay and lively, and in the highest spirits: the girls all very pleasing, and the young Walronds also. I do not reckon poor Cecil among the young people, as he is in sadly bad health. Mr. Walrond exceedingly pleasant, as he always is, as well as very estimable. Mrs. Hutchings (she was Miss Farquharson of Invercauld), remarkably good-natured and good-humoured, very lively and cheerful.

January 16th.

Truly wintry weather. The early days of the month were mild, so that on the 6th and 8th I heard a ringdove cooing as if it were spring, and saw the little yellow winter Aconite in blossom. But on the 10th it turned very cold, snow began on the 12th: that night there was a heavy fall and hard frost, and the 14th, 15th and 16th have been true *old-fashioned* winter days,—bright and clear, the ground covered with snow, and the frost strong and steady. Thermometer down to 10 deg. Fahrenheit in the night before last, and to 12 deg. last night.

January 17th.

The very long debate in Parliament on the Address (really on the question between the Irish Home-Rulers and the rest of Great Britain) came to a division the night before last; 57 Parnellites to 435. Eight English members voted in the minority. The Conservatives (except some who were absent) very properly voted with the Government, but several of them (with good reason) sharply censured the Ministry for allowing the disorders in Ireland to go on so long before making any attempt to check them.

The speeches, of which I have read the full report in *The Times*, are those of Sir Stafford Northcote, Mr. Gladstone, Mr. Gibson,* Lord Hartington and Mr. Parnell†—beside Lord Beaconsfield's, and Lord Granville's in the House of Lords. Both Sir Stafford Northcote's and Lord Hartington's I thought excellent, and indeed Mr. Gibson's also. It is some satisfaction to find the Ministry (and even Gladstone) at last confessing the melancholy and disgraceful state of Ireland, and the necessity of " extraordinary " measures to put it right. But I am afraid it will be long before they will be able to carry a tolerably strong measure through the house.

LETTER.

Barton, Bury St. Edmnnd's,
January 18th, 1881.

My Dear Edward,

I was very sorry to hear that you were so unwell, but I hope I may infer from your last

---

* Mr. Gibson was Attorney-General for Ireland in the late Conservative Ministry.

† Parnell's speech was a real curiosity.

1881. letter, that you are convalescent, and that with due
care and prudence you will soon be quite recovered.

Very sorry too, not to have your company here with
the pleasant party we had last week, which I think
you would have liked; but I am sure you were
quite right not to move from home in such severe
weather when you were not well; it would certainly
have done you harm; and now the weather is worse
than ever.

The lowest that the thermometer reached here,
was 8 degrees Fahrenheit, but at Ickworth, I hear,
it went down to 4 deg. To-day it is not so cold by
the thermometer, but there is a horrible, furious,
east wind blowing, which makes the cold worse than
ever to one's sensations.

(January 19th.)—I think yesterday was—as to the
weather—the very worst day that I can distinctly
remember, or very near it. The wind was positively
frightful, and the snow was drifted and whirled
about in a way that was quite extraordinary. A
tall young Fir tree (a Cephalonica), was torn up by
the roots and blown down right across the road in
the park, just in front of the fly in which Katharine
and Rosamond were coming from Bury, they had a
narrow escape. A huge elm tree was blown down
across the high road, opposite to Baldwin's
(formerly Paine's) farm, and smashed one of the
farm buildings, it is a great mercy it did not kill
anyone.

We were expecting a large party yesterday
evening; three of them—Barnardiston and his two

daughters—have not yet appeared, and we do not know what has become of them ; it is a mercy that all the rest arrived safe. No London letters or newspapers have yet arrived, and it is near 1 p.m. I have not yet been out, but I hear from Scott and Allan, that a great many trees have been blown down, especially firs and pines, but fortunately none of our particular favourites. I suppose the weather has been nearly as bad in London as here. I hope you will take great care of yourself while it continues so severe.

Fanny has not been quite well the last few days— owing I suppose, to the weather—but I trust it is not anything of consequence. I am quite well— bating eczema, which is but slight.

Arthur is to enter at Sandhurst on the 10th of next month, an important step in his life. I am glad to see in the papers that Gladstone has at last been roused to give a vigorous and just rebuke to the tactics of the Parnell gang; I wish he may carry a really strong and efficient measure for the repression of outrages. Parnell's speech on Monday night, I think was more *honest* than his first smooth and plausible one, as in the last he openly avowed as his object the separation of Ireland from Great Britain.

(Afternoon).—Barnardistons and the two girls who had been missing, arrived while we were at luncheon. They had been unable to reach Bury yesterday ; had passed the night at Melford, and were above two hours to-day, in coming by train

1881. from Melford to Bury. Scott told me this morning that each of the three entrances to this park was more or less blocked by fallen trees.

With much love from Fanny, believe me,

Ever your affectionate brother,

CHARLES J. F. BUNBURY.

JOURNAL.

January 20th.

The storm on the 18th was one of the most tremendous I ever remember. The fury of the wind was really terrific, and the appearance of the snow whirling and drifting in clouds before the gale which reduced it to dust, was most extraordinary. The storm seems to have been general almost all over Great Britain, and to have done immense damage, causing, I fear, great loss of life on our eastern coast. Here, there have been some narrow escapes. A fine, tall, young fir tree (of the variety Cephalonica) was blown down right across the road from our west lodge to the house, falling just in front of the fly in which Katharine was coming from the station. All the three principal entrances to the park (Scott tells me) were blocked up more or less by fallen trees. A huge old elm tree fell directly across the high road, opposite to the "Elms" farm, smashing the roof and upper story of one of the farm buildings. I have not been able to go out since, but Scott tells me there has been a great destruction of trees, especially in the grove between

the east approach and the wheelwright's yard, and 1881.
that between the arboretum and the north lodge.

Since writing the last pages, I have gone with
Fanny through the arboretum and the north grove.
The principal loss which I observe is the fall of a
cut-leaved Alder, Alnus glutinosa, var. laciniata, at
the south west corner of the new arboretum ("Sor-
cerer's paddock"*) and on the edge of the north
grove ; this is a great loss, for it was a beautiful tree
and a very fine one of its kind. But it is a great
comfort that none of my especial favourites have
suffered ; neither the Catalpa, nor the Tulip trees,
nor the great Magnolia acuminata, nor the Black
Walnuts, nor the Deodars, nor the Pinus excelsa,
nor the Picea Cephalonica in the arboretum, nor
the Cedars and Cypresses in the pleasure-ground.

———————

January 22nd.

Again a very lively party in the house from the
18th till to-day. Major and Lady Florence
Barnardiston and two of their daughters; M. and
Mme. Ernest de Bunsen and their second daughter;
Cissy and Emmy ; Katharine and Rosamond ;
Colonel and Mrs. Ives ; young Reginald Talbot
(Mrs. Ives's brother).

They arrived in the very midst of the tremendous storm, and it was a great mercy that none of
them were hurt. Katharine (as I have already
mentioned) had a narrow escape.

Barnardiston and his daughters, indeed, did not

---

* It was so called from a favourite horse of Sir T. C. Bunbury's being buried
there. I am told he was the sire of many of the horses in the Blues.

1881. arrive till the afternoon of the 19th, having been unable to get further than Melford.

In spite of the bad weather, the young people all seemed to be in excellent spirits, and were very merry. They danced on the nights of the 19th and 21st, being reinforced by some of our neighbouring friends ; and on the 20th they got up a very amusing and cleverly-acted entertainment, supposed to be Mrs. Jarley's exhibition of waxwork. Reginald Talbot was capital as Mrs. Jarley.

January 23rd.

This is the most severe winter I think, since 1860-61 ; certainly since '70-71. The thermometer in my garden (observed by the gardener at 9 a.m. each day) was down to 6 deg. Fahrenheit in the night of the 20th-21st, and to 4 deg. in the 21st-22nd, but in the Abbey grounds at Bury, this latter night, I am told it went down to 4 deg. *below* zero Fahrenheit.

January 24th.

The newspapers are full of particulars of the havoc cuased by the storm in all parts of the country, from the North of Scotland even to Penzance. I should think that such a combination of an extraordinary gale of wind with snow and intense cold, very rarely happens. Here the damage done is much greater than I at first understood. A tall spruce fir, growing close to the arboretum wall near the N.E. corner, was blown down, and in its fall beat down a great piece of the wall. A fine Wisteria, which grew on that part of the wall, and which had extended itself from thence to the fir tree, and climbed the latter even to its top, has been dragged down in the fall of its supporters, and, I fear, a great part of it is crushed. A tree growing at the upper end of the American garden, has been uprooted, and I fear it must have crushed a number of shrubs growing in the adjoining beds; but in this as in many other cases, the real amount of mischief cannot be ascertained till the fallen trees have been removed.

––––––

January 29th.

Received a very interesting letter from dear Rose Kingsley, describing the effects of what she truly calls "that really awful Tuesday" (the 18th). "Here (at Tachbrook) happily, we are sheltered, "and have lost no trees. But the shrieks of the "wind, the snapping of branches, the whirling twigs "and driving snow against the windows—which we "expected every moment would come in bodily— "made as frightful a scene as I ever remember." Her brother Grenville had gone early in the morning to Banbury, to a horse-fair, and they had heard nothing of him till 10 at night, when "a "brave post-office clerk appeared with a telegram "to say that the line was blocked, and Grenville "didn't know when he should get home, but had "wisely stayed at Banbury. He arrived late next "day. Round Banbury and Edgehill, 7 men have "lost their lives. The accounts of the suffering "among the London poor at the East end are "heart-rending. Even here there are hundreds of

1881. " men out of work. I have been quite perished
" with cold myself, and am longing for the enchant-
" ing sunshine which Mary (Mrs. Harrison) describes
" so vividly at Genoa. The snow has been far
" worse round Wormleighton than here, and many
" of the outlying villages have been without any
" bread for a week! The people at one forsaken
" place, called Claydon, were actually starving for
" want of food, and none could be taken to them, as
" all the roads were blocked."

February 1st.

The news from South Africa—which came on the
29th of last month—is very uncomfortable :— Sir
George Colley's force decidedly repulsed in their
attack on the position of the Transvaal Boers, with
the loss of several valuable officers. They have
now taken up a defensive position, where they will
wait for the arrival of reinforcements which will
soon join them. It is now evident that the Boers
are in earnest, and determined to fight ; and as they
are all skilful marksmen, accustomed from their
childhood to the use of fire-arms, and as the
country abounds in strong positions, they will
probably give our army a great deal of trouble.
The war appears to me a most unsatisfactory and
indeed deplorable one, and it is melancholy that our
brave fellows should be sacrificed in a contest of
which the result cannot (as far as I see) be either
honourable or really advantageous to our country.

February 4th. 1881.

My 72nd birthday.

Thanks, most humble and hearty, to Almighty God, for all his goodness to me, especially for allowing me to live to this age in good health and in the enjoyment of so many blessings,—above all, in the love and companionship of my inestimable wife.

LETTER.

Barton Hall, Bury St. Edmund's,
February 5th, 1881.

My Dear Katharine,

I thank you heartily for your kind letter and good wishes on my birthday. Seventy-two years old! I seem quite an antiquity; but seriously I am very thankful, and have great reason to be, for having been permitted to live to this age, in good health and in the enjoyment of so many blessings,— above all, having my wife beside me also in health. I am very much grieved at hearing that the Miss Lyells are all so ill, their being all laid up by illness at once, and unable to help one another, makes it the more sad. I do earnestly wish and hope that they all may soon recover; the account given us by Rosamond's letter this morning was not very comfortable, but I trust there will soon be an improvement. I hope they have good medical advice within reach—better than was accessible when Fanny and I were last at Kinnordy. We shall be very anxious for the next news of them.

1881.   I wish you could see a Dendrobium nobile which is now in blossom in the conservatory (brought from the hothouse),—it is really magnificent: I think it has fully 60 blossoms open at once. I have never before seen so fine a specimen of the plant. The weather is now pleasant, and there is a profusion of the little yellow aconites in blossom under the trees, and many snowdrops in the beds near the house.

Yesterday we took a walk to see the wrecks caused by the storm: but I am glad to say that, though there are strange gaps made in several parts of the groves, especially by the side of the approach from the East Lodge, and between the arboretum and the North Lodge, I do not find that any of my special favourites have suffered, except one of the Wisterias.

The servants had a dance last night, which seems to have been a great success; and this morning Fanny and I walked to see Norman, as I had missed him from the party, and was afraid he might be ill: but it was nothing worse than rheumatism.

Mrs. Praed came to luncheon, and spent much of the afternoon with Fanny; her son, who has been staying here some days with Arthur, is a very nice lad, well educated in every way.

I have forgotten to thank you, as I ought to have done, for the pretty piece of drapery which you have sent me for my chair or sofa. I thank you very much for your kind feeling.

With much love to Rosamond, believe me ever,

Your loving brother,

CHARLES J. F. BUNBURY.

JOURNAL.

I see in the newspapers the announcement of the 1881.
death of Thomas Carlyle. It had been expected for
several days, and he seems to have departed by
a gentle, gradual decay, without any positive illness.
He is a man who will have left a deep mark on
his generation, and will not soon be forgotten. I do
not think that his influence or his teaching was
entirely good or wholesome ; by no means entirely ;
the tendency of it is to an unqualified worship of
mere strength, of mere power and energy, which
I think decidedly unwholesome, and indeed dangerous. But he was certainly a most sincere, zealous,
honest-hearted man, and a very powerful writer.
His French Revolution is one of the most impressive
books I have ever read; I cannot read it without
a feeling of actual awe.

The life of John Sterling is a very agreeable book
—perhaps the most *agreeable* of Carlyle's writings,
and very remarkable, because he has made so
interesting a biography out of such slight materials.

I have not seen Carlyle for very many years ; but
for a time—about 40 years ago, if I remember right
—I had several opportunities of meeting him. It
was when old Mr. and Mrs. Buller, the father and
mother of Charles Buller, were living with their son
Reginald at his parsonage at Troston, a few miles
from this, Carlyle visited them there, and stayed
some time ; they brought him to see us at Barton,

1881. and I had a good deal of talk with him while walking about the grounds. I was very much impressed by the union of simplicity and solemnity in his manner and discourse; and I remember thinking that the English (or rather Scotch) that he *spoke*, was much better than that which he wrote in his books.

LETTER.

Barton,
February 7th, 1881.

My dear Edward,

Very many thanks for your very kind letter and good wishes on my birthday. I have indeed great reason to be thankful for enjoying at my time of life such good health and so much happiness. I cannot foresee what may be awaiting me, but whatever it may be, I can never be grateful enough for the blessings which have been granted to me. I am very sorry that you gain ground so slowly, and I am much afraid that this return of cold weather may throw you back. The weather to-day has been terrible—the wind almost as furious and the cold almost as severe as on the 18th of last month, though the snow has not as yet lain much. It is a most extraordinary season.

Two days ago the snowdrops and yellow aconites were opening in such numbers to the sunshine, they looked quite like spring; now we are thrown back into the depth of winter.

I am reading the Life of Lord Campbell, but

I must confess that I think Mr. Hayward's article on it in the *Quarterly* much more entertaining than the book itself. There is to my thinking rather a superabundance of letters, but I do not doubt that the book would be more entertaining to one who had a more practical acquaintance with a lawyer's life. The *Quarterly* article is *very* entertaining. What a bitter article that is on poor Kinglake!

I am much interested by what you tell me about Irish affairs; that is, about what is thought of them; and very glad to hear that those who know the country think that the Land League is losing ground, and that the agitation seems likely to collapse. It would be a great blessing if Parnell, Dillon and Co. would take themselves off to America and not come back. We have indeed great reason to be anxious about George, and I feel very much for Cissy, for I am afraid the Boers' war is a very dangerous one, as they seem to be quite in earnest.

With Fanny's love,

Believe me ever your affectionate brother,

CHARLES J. F. BUNBURY.

Barton Hall,
February 7th, 1881.

My dearest Emmie,

I thank you most heartily for your very kind letter and good wishes on my birthday, and also for the book you have been kind enough to send me. Your loving words and kind wishes are

1881. very dear to me, for I love you very much, of that be assured.

I can hardly at my age reasonably expect *many* more birthdays, but I trust that those which may remain to me may find me as well and as happy as I am now, and that the number of my friends may never be less.   It will be a great mercy if we all get safely through this winter, which is one of the most frightful that I ever remember.   To-day is almost as terrible as the 18th of last month,—at least the wind seems almost as furious, and though the snow does not yet fall heavily, there is a threatening of a good deal more.   I am afraid we shall lose several more good trees.   It is a most extraordinary season.

I am very thankful that aunt Fanny has shaken off her cold, and that I am well, as I trust you are.

I hope we shall soon have good news of dear George; one cannot help being very anxious about him, for I fear the Boers will make a hard fight and are formidable marksmen.   In my opinion our troops ought never to have been there.

Arthur goes to town the day after to-morrow, and to Sandhurst the next day ; of course he is looking forward with great eagerness to his entry into *the world*.

Give my best love to your dear Mother, with very many thanks for her loving letter, and thanks for the book.   I will write to her soon.

With much love from aunt Fanny, believe me—

Ever your loving uncle,

CHARLES J. F. BUNBURY.

JOURNAL.

Edward writes to me on the 4th (he has been out 1881. of health for some time, owing to the hurtful effects of the weather on his liver) :—

"It has been amusing, while dining at the Athen-
"æum day after day, to have the Members from the
"House of Commons dropping in with the news,
"especially of the strange proceedings of the last
"few days. I am very glad to find that the general
"opinion among those who are really acquainted
"with Ireland, is, that the Land League is losing
"strength and influence from day to day, and the
"whole movement tending to collapse. Tenants
"who have joined it under constraint are beginning
"to pay their rents; and altogether the agitators
"seem to be cowed as soon as they perceive that
"the Government is in earnest."

———

A very fine tall Virginian Juniper (Red Cedar) of the upright or pyramidal variety was blown down in the storm of the 7th; it is a loss. It is not uprooted, but the stem broken and twisted round at about 2ft. from the ground.

Dear Arthur left us, for London first, and next for Sandhurst, where he is to enter as a cadet.

———

A capital account of Arthur, who is now fairly
entered at Sandhurst.

The news from South Africa again very bad: Sir
G. Colley's last action with the Boers was evidently
not a victory, but a drawn battle at best, with a
lamentable loss of brave men. I fear he has been
rash, and (as is so often the case with our commanders), has under-rated his enemies. The
effect has of course been to encourage the Boers
and increase their resolution; and at the same time
it appears that a strong anti-English feeling has
been kindled among the Dutch in the old Cape
colony.

February 12th.

Again a heavy fall of snow, but without wind
most of it melted before night.

February 13th.

The news (in *The Globe)*, of the marriage of
Baroness Burdett Coutts to Mr. Ashmead Bartlett.

February 17th.

A letter from Katharine brought us the sad, but
not unexpected news of the death of Marianne
Lyell, the eldest of the three sisters who had
remained till now. She was in her 80th year, but
had seemed till very lately to enjoy such vigorous
health and so strong a constitution, that one could

hardly suppose her so old, or think of her death as 1881.
near. She was an excellent woman, and was a
most valuable support to her brother Charles in his
desolation after the death of Mary. The other two
sisters are thought likely to recover, and I sincerely
hope that both may, for if either were to die, the
situation of the lonely survivor would be very
melancholy.

———

February 24th.

Two Weymouth pines, which grew between the
arboretum and the north approach, and were blown
down on the 18th of last month, have been
measured,—one was 80 feet high, the other 84 feet ;
both were planted by our father.

———

LETTER.

Barton,
February 24th, 1881.

My Dear Edward,

I heartily hope that your health has improved since we last heard from you. I am very
thankful to be able to say that both of us continue
free from any ailment of consequence.

You may possibly not have yet heard that our
nephew George has landed at Portsmouth safe and
sound, and Sarah, who saw him on his arrival says
he is looking very well and very handsome. I think

1881. he does not yet know whither he will be ordered—I hope not to Ireland.

Minnie Napier is with us, to our great comfort, and except her I have seen very few people this month: but Fanny and she go out driving; I do not think I have been outside our grounds since the middle of January.

I am reading Wallace's new book, "Island Life," and find it very difficult,—at least the astronomico-geological part, in which I am nearly stuck fast.

I hope we shall soon have a good account of your health. With Fanny's best love, believe me ever,

Your affectionate brother,

CHARLES J. F. BUNBURY.

JOURNAL.

February 27th.

Poor old John Brand has at last been released from his misery; he departed yesterday, and I can only rejoice, and heartily, that his long and terrible sufferings are at an end. His malady was a cancer in the face, and he has endured this dreadful disease for a frightful length of time; it is indeed wonderful how long it has preyed upon him before destroying him. He was for many years the principal labourer in the garden here; my father had a very high esteem for him, and used to compare him to Adam Bede.

News from South Africa more and more deplorable. Our troops again beaten by the Boers, with lamentable slaughter, and it is believed that Sir George Colley himself has been killed. What a number of unhappy wives and parents will be in miserable suspense till the names of the killed and wounded are known!

———

The names of the killed and wounded officers in the disastrous fight on the "Majuba" mountain, or "Spitzkop," are now published in *The Globe:* and they rather relieve one's mind, for the killed or severely wounded are not so numerous as was at first supposed. Sir George Colley himself is indeed slain, and young Maude: but Major Hay and Commander Romilly, who had been supposed dead, are returned as "slightly wounded," and one or two others who were missing seem to have re-appeared. Many particulars of the action are also published, having been witnessed by special correspondents of newspapers, who were captured in the rout by the Boers. Still it does not seem clear how the disaster happened — such a sudden reverse of fortune, for the troops seem to have been holding their ground successfully, and to have even been confident of victory, till the enemy made their final charge. I suppose that the Boers were very superior in number, that the position of our troops (though strong) was too extensive to be

1881. thoroughly occupied, and that the enemy broke in at some weak point. It is clear that the Boers fought splendidly, and that they are formidable enemies. It is lamentable that the blood of so many good and brave men, on both sides, should be shed in a useless war. It would be rash and wrong, before we know the circumstances more thoroughly, to blame poor Sir George Colley, or impute rashness to him, although the first look of the case may incline one to such a judgment.

— — — — —

March 4th.

We have now, from the newspaper correspondence, a great many particulars as to the terrible disaster on the Majuba mountain, but I do not yet clearly understand what was the cause of it, or how it happened. In particular, it does not yet appear what numbers the enemy had, though I presume that they must have outnumbered us very much,—nor whether their seemingly sudden success at last was owing to their receiving a sudden large reinforcement, or to their discovery of a weakness in our position, or to an unaccountable panic which may have seized our men,—for such things have been known to happen. It seems that there were some mistakes or uncertainties in the first lists of killed and wounded : for poor Commander Romilly who was reported "slightly wounded," is now said to have died of his wounds, and some who were missing have reappeared. Colonel Stuart, who was supposed to have been killed, is, it seems, a prisoner, unwounded : I am very glad, not that I 1881. know him personally, but his wife (who was the widow of Sir Henry Tombs) is a great friend of Cecilia, and was so of my dear brother Henry.

———

The Irish Coercion (or Protection) Bill received the Royal assent on the 2nd, after being more furiously and desperately opposed in the House of Commons, and taking up more time (I suppose), than almost any measure ever was before. I hope it will be vigorously enforced, and will have good effect in Ireland. I think the Government ought to have brought forward some such measure earlier : but certainly they have since pushed it forward with spirit and steadiness. But the Arms Bill, which is a sequel to it, has only just now got into committee, and is not likely to become law for a long time yet : and then there is the Land Bill. Altogether, Ireland seems likely this year to monopolize the attention of the House of Commons.

———

Have just heard the sad news of the death of James Spedding, by a deplorable accident—run over by a cab in the street. I am very sorry, though I had seen hardly anything of him for many years past. But he was a man very well deserving of respect. My acquaintance with him began a long time ago, immediately after my entrance at Trinity

1881. College, Cambridge, in the October term of '29. He was my senior in *academical age* by two years, for he took his degree in '31, and I, if I had stayed on, could not have taken mine before '33; but in years he may not have been older than I. Even when an undergraduate he was very old-looking (singularly *bald*), and very grave and sedate in manner; and I believe he was even then very studious, though he did not take high honours in the special studies of the place. After I left Cambridge, we did not often meet, and of late years communication between us has been very unfrequent.

Spedding's "Life of Bacon," and his edition of Bacon's works, have given him a high and permanent place in literature.

Yesterday's *Times* gave us the melancholy news of the death of our friend Lord Hanmer. There are no particulars, but it took place at a distance from his home, and I therefore conclude it was sudden; which is not surprising, as he has repeatedly of late years, had very severe attacks of gout; and last spring he was so ill, that most who saw him thought he could not last long. Yet he rallied, and appeared to be in tolerable health when we were with him at Bettisfield in September. He was a most amiable, kind, warm-hearted man; a most cordial and zealous friend, as Fanny and I have experienced for many years past. Since we have had a house in

London we have seen him there frequently every 1881. spring, and have become quite intimate; he has appeared to take a great liking to Fanny, and they have kept up a pretty active correspondence. She had a letter from him little more than a month ago, written in a cheerful spirit, though showing that he felt the depressing effect of his loneliness and of the ungenial season. He was of a very social disposition, and although he did not lack mental resources, seemed ill fitted to endure a solitary life. The death of his wife (just about a year ago), was a terrible blow to him, yet he did not shrink long into solitude, but sought society again within a short time, in spite of his own ill health. He was not only a good and amiable man, but one of a culti-vated mind and of considerable acquirements.

March 14th.

The news of the horrible murder of the Emperor of Russia—most startling indeed. Truly we live in most extraordinary times : all the moral foundations of society seem to be giving way. I hardly remem-ber in my time an event so shocking as this murder. Yet perhaps I hardly ought to say so, for Fieschi's " infernal machine " and the " Orsini bombs " may have been equally atrocious in design, and (in Fieschi's case at least) caused a greater destruction of life, though they failed of their especial object.

March 17th.

Fanny has had a letter from one of the Miss

1881. Kenyons, Lord Hanmer's nieces, giving us some information about the death of our friend.   He was staying with his friend, Lady Harriet Warde, at Knotley Hall in Kent, and had seemed to be very fairly well, but caught an accidental cold, which attacked his lungs. and he sank rapidly.   His last hours it is said, were very calm and happy, and he expressed his gratitude that he was dying amidst such kind friends, instead of in the loneliness of Bettisfield.

————

March 24th.

The news yesterday and to-day from South Africa is very good.   Peace seems to be *all but* concluded between us and the Transvaal Boers, commissioners being appointed to settle definitively the details of the treaty on each side.   The commissioners are—Sir Evelyn Wood, Sir Hercules Robinson, and the Chief Justice of the Cape on our side ; and on the other, the " Triumvirs," who appear for the present to govern the Transvaal. The essential substance of the conditions appear to be :—that the people of the Transvaal are to enjoy complete *autonomy* as to their internal affairs, but will acknowledge the British flag, and submit to the authority of this country in all their foreign concerns —that our troops are to remain for the present in their country, to guard against any disturbances which may arise ; and that our troops are not to be withdrawn until after the armed Boers have dispersed to their homes.

I am heartily glad that no more lives of brave men are to be sacrificed in a war which we ought never to have entered into; but at the same time, I cannot deny that our concessions in this case— made after defeat—may possibly encourage other malcontents in South Africa to set us at defiance. and that future wars in that country may arise out of this one.

I have no doubt that many of our young officers, who are burning to distinguish themselves in this field will be bitterly disappointed by the news of peace.

————

March 25th.

Clements Markham told me that the India-rubber from the province of Ceará in Brazil, has for some time been known in trade, as the second best in quality, inferior only to that of Pará, but till very lately it was not known from what plant it was obtained; that it is now ascertained to be a species of *Jatropha* (of the Euphorbiaceæ); and he hopes and believes it will soon be introduced into India. As Ceará is comparatively *(for* Brazil) a dry region, it is hoped that this Jatropha may be cultivated in some of the dryer parts of India.

In the moist climates of that country, the Pará India-rubber plant is already cultivated with success.

The uses of Caoutchouc are now so multifarious and the demand so enormous, that every available source of it is important.—(Clements Markham).

1881.   Clements Markham tells me that a steamer has now made the voyage to the Cape in nineteen days, the shortest passage hitherto known, the regular time of the voyage by this line of packets is twenty-one days. When I went out to the Cape, in '38, in a sailing ship, the voyage lasted sixty-six days !

Clements Markham has been for a good while engaged in writing a history of the great family of De Vere, and (he tells me) finds the subject extremely interesting; but the business of the Geographical Society, and especially the editing of its Journal, takes up much of his time. He has however got together and arranged the materials for his history, so that he will have only the labour of composition.

March 27th.

Very agreeable society in our house from the 19th to this day :—Courtenay and Lady Muriel Boyle, Leopold and Lady Mary Powys, the Clements Markhams, Edward, Ellen and Constance Wilson, Ada Newten, Gertrude Waddington : and now Mrs. Wilson and Agnes. The Boyles, Powyses and Markhams, I have often mentioned before, and need not repeat their praises. The Markhams' visit was too short—they were obliged to leave us yesterday. Alfred Newton (the Professor) gave us a still more flying visit of only one day.

Alfred Newton acknowledges that his edition of Yarrell goes on very slowly ; he says that the quantity of new matter accumulated since the first edition is enormous, and the difficulty of not only selecting and arranging, but compressing it within a reasonable space, very great. In particular, he has a great deal of curious new matter about the cuckoo.

— ——

Helen Ellice was married on Monday, the 28th to Mr. Percy Lambart. Minnie went from us to London on the 26th to attend the wedding (which seems to have been very satisfactory), and returned yesterday. Helen is very good, and a very pleasant girl, and I heartily wish her happiness in marriage.

——— —

Dear charming Lady Muriel Boyle left us this morning; I was very sorry to part with her. She is truly charming. Her husband left us the day before yesterday, and expected to return yesterday evening, but official business called him to London. We are now alone, for I consider dear Minnie as one of ourselves, not a visitor.

I am afraid that the peace with the Transvaal Boers, seems likely to be not at all conclusive or lasting.

— — — —

By the Census of 1881 ;—
Population of parish of Great Barton in 1871—878.

1881.   Pop. of same in 1881—817.
                            Decrease of 61.
                      In 1871.
                            Males in Barton—412.
                            Females ,,   ,,   466.
                      In 1881.
                            Males   ,,   ,,   404.
                            Females   ,,   ,,   413.

                                                    April 14th.

Scott having returned yesterday from Mildenhall,
where he had had two very hard day's work in the
Rent Audit, made his report to me this morning.
It was not more unfavourable than I had expected.

                                                    April 20th.

A telegraphic message to Bury, yesterday morning, brought us the news of Lord Beaconsfield's
death.   His constitution struggled for a longer time
than might at his age have been expected, against
the disorder: and indeed it seems that the actual
disorder had been overcome, and that he died of
want of strength. Though I did not know him
personally, I am very sorry for his death, and I
think that the feeling of regret will be pretty general
among candid and liberal-minded men, whatever
they may think of his special political career.

I cannot but admire a man, who, under such
disadvantages and difficulties of race and position,
and at first of personal prejudice, raised himself to

the most commanding place in the State by sheer energy, determination and genius. There have been many things in his political course which I have not approved : but I do not know of anything which should prevent one from feeling a true respect for him, both as a statesman and as a private man.

To his political party the loss is enormous, and probably irreparable. The greatest and most remarkable achievement of his political life, I should suppose, was the recreation of the Conservative party, after Peel had shattered it to pieces by the repeal of the Corn Laws, and the leading it and keeping it together under such discouraging circumstances.

Neither the late Lord Derby, nor Lord Salisbury, nor Sir Stafford Northcote, I take it, could have accomplished this without Disraeli.

Mr. Sanford tells me that he finds the lesser spotted woodpecker to be not uncommon about his home, and believes it to be not generally rare in England, but it is very shy and not easily observed : he has ascertained that it is the bird which makes a curious noise in the spring, as if a tree-trunk were splitting, and that the noise is made by a very rapid succession of taps on the tree. He spoke of the various strange noises heard at night in the Australian forests; above all, the dreadful sounds uttered by that curious bird the *podargus*— the most painful and distressing sounds imaginable, like the screams of a woman under the most violent ill-treatment. Wonderful noises uttered by a very

1880. small kind of frog or toad which lives in holes in
very dry ground, far from water. Besides the
podargi, a species of true caprimulgus in Western
Australia; Mr. Sanford kept one alive for six
months, and it grew very tame : he fed it partly on
moths and beetles, partly on small morsels of raw
meat. Mr. Sanford was much struck with the
differences as to their animals, between the different
portions of Australia,—the north and south as well
as the east and west. He adopts Wallace's theory,
that that continent was formed by the union of
originally separate islands, but differs from Wallace
in believing that they were more than two. He
mentioned an extremely venomous snake he saw in
Western Australia—velvet black, with an absolutely
scarlet belly :—also a very small and very beautiful
kind, variegated with scarlet, black and white, of
which he kept several alive, but could never
ascertain positively whether they were venomous or
not. Mr. Sanford told me (speaking of the great
antiquity of some forms of life, both vegetable and
animal) that he has specimens of fossil wood, *not*
Coniferous, from the Lower Coal formation in the
Bristol Coal field :—wood of the normal Dico-
tyledonous structure, without the *disks* of the
Coniferous tissue, but with the vessels marked with
spiral lines. He said that the structure of the shell
of Echinoderms is very peculiar, and easily recog-
nizable even in the smallest fragment, and that this
structure he has discovered in limestone pebbles
contained in a conglomerate of Laurentian age.
Now the Laurentian is the oldest series of rocks

yet known to us, and if a conglomerate of that age 1881.
contains fragments of organic bodies (which must
have been derived from a still older formation) this
carries back the age of organic bodies to a
wonderful antiquity. And Echinoderms are by no
means in one of the lowest grades of known
creatures.

April 26th.

Another very agreeable party in our house since
the 20th :—Mr. Sanford, with his daughters Ethel
and Blanche and his son Ayshford, Lady Head and
her daughter Amabel, Ellen and Constance Wilson,
Edward, Harry, George and Arthur, and (from
Saturday to Monday), young Lothair de Bunsen.
A very merry party of young people and charming
girls. Ethel (Ettie) Sanford is exquisitely lovely :
her sister very pretty, and well deserving of her
name of Blanche ; the two Wilson girls very
pleasing ; Amabel Head very sensible, intelligent
and agreeable. Mr. Sanford is a very remarkable
man, and I think it a great pity he is not more
known ; the extent, variety, and accuracy of his
knowledge are really astonishing ; in these qualities
and also in the eagerness and impetuosity of his
character and manners, he reminds me sometimes
of Charles Lyell, and sometimes of Charles
Kingsley. Mr. Sanford spoke with enthusiasm (in
which I fully concur with him) of the beauty of
Madeira : and much also of the miserable condition
of its inhabitants at the time of his visit.

LETTER.

Barton, Bury St. Edmnnd's,
April 27th, 1881.

My dear Katharine,

1881.   The title of the little book on Alpine
plants which you ask about, is "Die Alpen Pflauzen
"Deutschlands und der Schweiz, in colorirten
"Abbildungen, von J. C. Weber, mit systematich
"geordnetem. Text von Dr. C.A. Kranz. München,
"1872."

The figures are in general pretty good, characteristic, *rough* likenesses, but by no means scientifically
accurate.   I do not however know any other book
with figures of the plants of the Alps of a convenient
portable size.   The book I should take if I were
going thither, would be — *Koch*, Synopsis Floræ
Germaniceæ et Helveticæ, Frankfort, 1837; but
I am afraid it is a little rare and difficult to meet
with.   It is in Latin, but *botanical* Latin, which
anybody can read.

De Candolle's Flore Française is an admirable book,
but it is in six 8vo. volumes.

I was just intending to write to you as soon as our
party should have broken up, but you have antici-
pated me, and I thank you very much for your
letter.   I was delighted to see *your book* advertised
as one of Murray's "forthcoming," and hope I
shall live to read it.   I admire your industry.   My
Botanical Fragments are by no means so near completion.

I hope you will make out your expedition to
Auvergne satisfactorily, have good weather, and
enjoy it much.

If you could pick up any mosses or lichens for me
in that country, I should be very much obliged
to you.

I have not time to write more at present.

Believe me ever, your loving brother,

CHARLES J. F. BUNBURY.

Thanks for lending us Mr. Horner's letter on
Auvergne, which is very interesting. Fanny will
copy it, and return it as soon as she has done so.

JOURNAL.

April 28th.

The dimensions of two white willows which were
destroyed in the great storm of the 18th of January
last (one completely uprooted, the other so much
shaken that it has been taken down as dangerous) :

One near wheelwright's shop—Height 86 feet to
top of boughs ; girth, 12 feet 9 inches at 3 feet from
ground.

One near east approach :—Height, 84 feet ; girth
10 feet 9 inches.

These are much beyond the dimensions of any
Willow mentioned in Loudon's arboretum.

LETTER.

Barton, Bury St. Edmund's.
May 6th, 1881.

My Dear Katharine,

1881.

Very many thanks for your very agreeable letter from *the Lizard*. I had no notion that there was a hotel there; it all looked so still and solitary in my time. I delight to hear of Cornwall, my remembrance is still so vivid of my tour there in '41, and my visit to Sir Charles Lemons, which I always look back on with special pleasure. Though it is close upon forty years ago (forty years next August) my impression of Kynance Cove, and St. Michael's Mount, and the Land's End, is still as fresh and vivid as can be. It *was I* who went with Sir Charles Lemon to visit Mr. Fox, near Falmouth; I remember it perfectly.

It is curious that only two days ago, I was talking with Miss Doyle about the *Miss Sterlings*, who are particular friends of hers, and now comes your letter with such a pleasant account of them and of their home. I am glad to hear of them again, as I have an agreeable remembrance of them (or rather of Julia, for I think I hardly knew the other sister) and of their father, though my acquaintance with him was only of three days. The bit of leafy shoot you have sent me looks most like *Myrsiphyllum asparagoides*, which is a South African plant, but it may possibly be *Ruscus* androgynus, which is of Madeira. The flower appears to be an *Abutilon*, but I do not know the species.

(*May 7th*).   Here also the weather has been most
beautiful these two days—quite delicious;   and I
have been fully feeling that "vernal delight" which
I enjoy every year so strongly at this time whenever
the season is propitious, and which I do not find at
all spoiled by age.

The profuse beauty of the spring flowers, both in
the garden and the fields, the tender colouring of
the young leaves, the songs of the birds, the revival
of life all around; these are pleasures which never
pall upon me.   It is true there comes every now and
then, amidst all this beauty, the touch of melancholy
remembrance, when we "turn from all she brought
"to those she could not bring," to those whom we
so long to have again sharing our enjoyment.

With much love to dear Rosamond, and to
yourself, from both of us, believe me,

Ever your loving brother,

CHARLES J. F. BUNBURY.

JOURNAL.

May 7th

Again a very agreeable company in this house,
departing to-day: Lord and Lady and Miss
Tollemache, Sir Francis and Miss Doyle, the
Montgomeries, John Hervey, and (for parts of the
time) the two Miss Bevans, Agnes Wilson and Mrs.
Arthur Wilson.—besides various dinner guests.

Sir Francis Doyle is a very entertaining and
very agreeable man, with a wonderful fund of

1881. anecdote and personal history, which he tells with a
great deal of humour; and besides this, he has a
great deal of real, sound knowledge.

Lady Tollemache is a sister of my brother Henry's
friend, Colonel Duff; she is very clever, has a
remarkably cultivated mind, awake to a great variety
of interesting lines of thought and study — has
agreeable manners, and (I think) great earnestness
of character.

Lord Tollemache, a very good and valuable
country gentleman, wonderfully active for his age
(76).

Mr. Montgomerie, very amusing; Mrs. Montgomerie very pretty and very pleasant.

_____

May 8th.

Most delightful weather the last three days;—
perfect spring weather, such as May ought to be,
according to the poets; bright sunshine, soft air,
light westerly breezes, warmth, genial but not
oppressive; in short, perfectly delicious.

The trees and shrubs, and plants of the gardens
and fields, long kept back by the east winds, are
now rushing into leaf and flower with exceeding
rapidity, so that every day brings out something
new. I am enjoying anew that "vernal delight"
for which I have so often felt reason to be
thankful; thankful enough, indeed, I can never be,
for the power of enjoying these delights so keenly,
and in the charming company of my beloved wife.
I hardly think that I ever enjoyed the beauties of
spring more heartily than in the last few years.   I

need only repeat what I wrote on the 14th May 1881.
last.

All the " periodical phenomena," the leafing and
flowering of trees, and the appearance of migratory
birds, are very late this year.

———

I have had a charming letter from dear Mrs.
Kingsley, but am sorry to learn from it that she has
been unwell. She writes beautifully, with true and
deep feeling, and the eloquence proceeding from a
beautiful mind, on the delight she receives from the
spring, and the ideas of resurrection and immortality which it awakens; but I have not time at
present to copy out this part of her letter.

She tells me that Rose is just returned from "two
"charming months in Italy," that Maurice is
prospering in Mexico; and that Mary (Harrison)
and her husband are come home after a delightful
winter (on the Riviera).

She also says that she has read " with deep pain
" Carlyle's ' Reminiscences,' a publication I deplore
" for his own sake, poor dear old man—for he has
" done *himself* irreparable injury—exhibiting only
" the morose and dyspeptic side of his character and
" mind. It seems to me a most unwise publication."

There are articles on Carlyle's " Reminiscences."
in both *The Edinburgh* and *Quarterly* of April. That
in the *Edinburgh* (which Edward tells me is by
Mr. Henry Reeve) is to my thinking, very fair and
just, and altogether very good ;—The *Quarterly*
article which is said to be (and evidently is) by

1881. Hayward, is much too harsh, too sweepingly severe, indeed bitter.

I have also lately had a very agreeable letter from Katharine, who has been making a pleasant, little tour in Cornwall, visiting Falmouth, Penzance, the Lizard and the Land's End.

Katharine describes her visit to the Miss Sterlings at their house near Falmouth—"which is an Eagle's " nest perched above a beautiful little bay, and such " a charming house, with every comfort and refined " luxury in it. They delight in flowers, and their " garden, formed in terraces, is full of beautiful " shrubs and flowers, not to mention trees, which " this climate enables them to have in such " abundance.'

The Miss Sterlings are daughters of John Sterling, whose Life by Carlyle is one of the most agreeable works of that writer. I made acquaintance with John Sterling in 1841, when we both were guests of Sir Charles Lemon, at Carclew; we had much conversation during the three or four days that we were in the same house, and I thought Sterling a very interesting man. I never saw him again. His writings were few and very disappointing. Long afterwards, less than twenty years ago, I met Julia Sterling, the elder of his daughters, several times, when she was living with her uncle, Mr. Maurice; she was then very handsome and very agreeable.

She (Katharine) also visited Miss Fox and her beautiful garden at Penjerick. Old Mr. Fox died only a few years ago; he took an interest in bringing every exotic tree that he could acclimatize, and the

variety is wonderful. The views of the sea, and 1881.
the very rich foliage, and the brilliant blossoms of
rhododendrons, make it a perfect paradise. The
variety of trees and shrubs is infinite, and Miss
Fox delights in *biology* in every shape.

LETTER.

Barton Hall, Bury St. Edmnnd's.
May 21st, 1881.

My Dear Katharine,

I cannot find anywhere in the Memoir of
Mrs. Somerville, a clear and precise statement of
how far the " Mechanism of the Heavens " was
original ; but it clearly was not a mere translation.
Sir John Herschel writes to her :—" What a pity
" that La Place has not lived to see this illustration
" of his great work !" And, as far as I can make
out, it was intended as an *illustration* of that work—
intended to throw light upon it, to make it somewhat intelligible to those who had a partial knowledge of the subject. But I must ask you to excuse
me from saying more, as I have a great deal to do
in the very few days that remain before we go to
town. I am very sorry that we arrive just as you
are going away, but I hope you will very much
enjoy your visit to Auvergne. I am very unwilling
to leave Barton at this season. I never saw it more
beautiful.

Ever your loving brother,
CHARLES J. F. BUNBURY.

JOURNAL.

1881.

48, Eaton Place.

We arrived in town yesterday, and are here, I suppose, for "the season." I leave Barton very unwillingly at this time of year, when it is most beautiful, and the effort has been the greater, because the weather from the 20th to yesterday was delicious—most enjoyable in every way. The lilacs, laburnums, hawthorns, and all the various kinds of fruit trees, are in the most profuse beauty of blossom, and above all, the Chinese apple tree, Pyrus spectabilis, which is capricious as to flowering, and has not done well for several years past, is now exquisitely beautiful.

May 29th.

We have fortunately come to London in time to have a glimpse of our dear Arthur Herveys. The Bishop came to luncheon with us on the 27th, and was delightful; and yesterday we found Lady Arthur at home in the afternoon, and had a very pleasant chat with her. Both are looking very well, and *he*, indeed, wonderfully young for his years.

May 30th.

The 37th anniversary of our happy wedding day —a blessed day indeed for me. I can never be sufficiently thankful to Almighty God, that after the lapse of so many years, I am still permitted to see my beloved wife by my side, and both in good health.

This was a most beautiful day. We drove in Battersea Park, which is delightful with its tender green young foliage, its lilacs and laburnums, and red flowered thorns and horse chesnuts in full beauty.

At dinner at Lord Tollemache's we met Sir Joseph Fayrer, celebrated for his knowledge of Indian serpents, and of Indian diseases. I had some talk with him about the climate and productions of that country, but he did not tell me much that I had not already known from Hooker.

———

June 1st.

A splendid summer day.

A visit from our dear old friend Mrs. Mills; she is looking well, seemed in good spirits, and was very agreeable.

We went to Mrs. Horton's afternoon party, in their handsome house at the corner of Grosvenor Place: it was a large party, yet not crowded, and I was quite surprised at the great number of acquaintances whom I recognized there—in fact almost every one whom I know who is at present in London : and many I was very glad to see.

———

June 3rd.

We had a little dinner party in compliment to

1881. Helen Ellice and her husband (Mrs. and Mr. Percy Lambart)—besides them there were Minnie, Cissy, Emmie, Major Heneage Legge, and Edward. It was pleasant. Helen is rather a favourite of mine : and her husband made an agreeable impression on us.

_____

June 4th.

I looked into the Flower Show at the Horticultural Society's gardens :—a very beautiful display of Orchids and Azaleas, with some fine Aroids and Ferns. The Orchids really superb, particularly the Cattleyas, Laelias, Dendrobiums and Odontoglossums. Noticed a very small but very rare and curious Fern — Actinopteris radiata — remarkably like a miniature Fan-Palm. The Actinopteris, though very rare in collections in Europe, appears to be scattered over a wide extent of country in tropical and sub-tropical Asia and Africa.

_____

June 6th.

Weather — all last week — most beautiful and delightful, perfect summer, very hot in the middle of the days, but not sultry even in London. On the 1st, as Scott wrote to me, the thermometer at Barton, at noon, was at 81 deg. in the shade. The rain has now come : yesterday was very wet, and this day showery.

It is strange that I have omitted to mention—in my journal—Montagu MacMurdo's elevation to be K. C. B. Certainly he deserved it, and ought to have had it long ago.

We dined yesterday with the Leonard Lyells. A very pleasant party:—Mr. Lowell (the American Minister), Mr. Rogers, Miss Peabody, Miss Elliott, Mr. Hughes ("Tom Brown,") Sir George and Lady Young, Emma Stirling, Mr. Sedley Taylor.

Miss Peabody, a young American Lady, is very pretty, very clever, highly cultivated, very lively and agreeable: quite enthusiastic for Kingsley, of whom, and of whose Life she talked much and well; she knew my name from that Life, and was eager to hear all I could tell her about him, saying that she envied those who had known him:—so we sympathized thoroughly. Miss Elliott* also very agreeable. Mr. Hughes (whom it is strange I had never known before)—very pleasant.

We attended the wedding of Miss Milnes (daughter of my old friend Monckton Milnes, Lord Houghton), to Mr. Fitzgerald), at St. Peters, Eaton Square, and afterwards went to the wedding *tea party* given at Mrs. Thornhill's house in Bruton Street. — An immense crowd. The wedding presents displayed—magnificent.

Montagu MacMurdo, talking of our deplorable defeat by the Boers at the Majuba mountain, and attributes it altogether to General Colley's errors, and not to the fault of the soldiers. The top of the mountain, he says, is somewhat hollowed or basin

---

* Daughter of the Dean of Bristol.

1881. shaped—probably the remains of an ancient crater.
General Colley kept the whole of his reserves
inactive for several hours within this hollow, where
they could see nothing; he kept up no communication with the fighting line, which was
engaged with the Boers far below on the hill side;
the Boers, finding part of the edge of the summit
undefended, came over it, and poured in on the
reserve, who, having been in ignorance of what had
been going on outside of their *crater*, were taken by
surprise, and a complete panic was the consequence.
The Boers, Montagu says, were actually inferior in
number to the troops occupying the position.—And
yet the regiments, which were so wretchedly beaten
here, were the very same which, under Sir F.
Roberts, had performed the famous march from
Caubul to Candahar, and gained a complete victory
over the Afghans.

Speaking of enlistment for long or short term
of service in the army, MacMurdo said he had no
doubt that the long service would be preferable,
but, that the short is inevitable, for men cannot now
be got to enlist for the longer terms.

_____

June 13th.

We (Fanny, Mimi MacMurdo, and I), visited the
Botanic Gardens in the Regents Park. They are
in great beauty. In the great conservatory, there
are very fine specimens of many Palms, Tree-ferns,
Aroids, and Conifers ; as Seaforthia elegans, Sabal
umbraculifera, Dicksonia Antarctica, Araucaria ex-

celsa, Cunninghamii, and Bidwillii. Also the beau- 1881.
tiful climber Petrea volubilis in profuse blossom.
In the "Economic" house, a Mango tree very
healthy and flourishing, with its peculiarly dark and
glossy leaves, and with fully formed though unripe
fruit.

———

Miss Elliot says, that Mr. Froude is quite surprised at the anger which has been excited by
the publication of Carlyle's Reminiscences. He
has accustomed his mind to such a subservience to
Carlyle's, as not to conceive that any one could
have a right to be displeased with what Carlyle
approved. She feels sure, not only that Carlyle
intended his Reminiscences to be published, but
that they were, if not printed, at least completely
prepared for publication, before his death. Miss
Elliot remembers a dinner-party at which she met
both Carlyle and Froude at the time when there was
much excitement about Governor Eyre and the
negro insurrection in Jamaica. Carlyle, as was to
be expected from him, spoke very strongly in favour
of Eyre; and upon *her* saying something for the
cause of the negroes, he replied : — "It is of no
"use your trying to make those *white* whom God
"Almighty has made black." Dean Milman, who
was present on the occasion, but (on account of his
deafness), had not heard what Carlyle said, remarked, on its being repeated to him, that it would
be unreasonable to object to *white-washing*, when

1881. those who had white-washed Henry the Eighth and Frederick the Second were at the table."

Miss Elliot, Minnie Boileau, and Lady Hoste came to luncheon : Miss Elliot very agreeable.

We dined with the Charles Hoares; had a pleasant evening. I sat between dear Kate, who is always charming, and Mrs. Pascoe Glyn, who is very agreeable.

———

June 18th.

I was very sorry to see in the newspaper, the death of Professor Rolleston of Oxford, at the age of only   He is a great loss to science, to zoology and physiology in particular; a most serious and heavy loss, I should suppose to the cause of Natural History at Oxford; for he was a man not only of great talent and knowledge, but of immense zeal and energy; the sort of man who does more to promote knowledge than a multitude of ordinary teachers. We met Professor Rolleston, and made his acquaintance through the Augustus Vernon Harcourts, when we were at Oxford last year.

He showed us part of the new Natural History Museum there, and we were much struck with the clearness, spirit and animation of his exposition and remarks, as well as with the liveliness and copious-ness of his general conversation.

———

June 20th.

Dear Rose Kingsley arrived to pass some days

with us, to my great delight. She gives a very good 1881.
account of her mother's state of health.

We (Fanny and I), looked into the Royal
Academy Exhibition, and spent an hour there ; it
was inconveniently crowded, but we noted a few
things, in particular, we were very much impressed
by *Long's* " Diana or Christ."—(A Christian maiden
urged by her lover and friends to burn the one
grain of incense which would free her from perse-
cution). It is certainly the finest modern picture I
have seen for many years—except perhaps *Gustave
Doré's* " Christ leaving the Prætorium." We
admired also two delightful river views (on the
Thames, I suppose) by *Vicat Cole.*

June 23rd.

Talking of the projected tunnel under the Straits
of Dover, and the controversy which has arisen in
The Times, as to facilities thereby afforded for a
French invasion of England, I find MacMurdo
thinks the apprehensions on this score are ground-
less, as the mouth of the tunnel could be so easily
and speedily closed by an explosion of dynamite.
This however seems to assume that our Government would be watchful, and on the alert, which
(if the Government were like the present) might
perhaps be a rash assumption.

June 24th.

The Hooker's garden party at Kew. The day

1881. beautiful (an unusual piece of good fortune for those
parties) ;—the grounds in full beauty ; a great many
of our acquaintances.

————.

I saw the *comet* last night, very distinct and
well defined, though very small in comparison with
that of '58; it was, when I saw it (from 11 to near 12
p.m.) low in the sky, in appearance not far above
the chimney tops of the houses opposite to us on
the N. side of the street; the tail (very distinct)
pointing almost vertically upwards.

Our dinner party; the Louis Mallets, the Mac-
Murdos (3), the Leonard Lyells, John Herbert,
May Egerton, Mr. Arthur Denman, Cecil, Lina
Bruce ; besides Rose Kingsley and Arthur
MacMurdo, who are staying in the house. A very
pleasant party.

——— ———

Dear Rose Kingsley left us to return to her
mother, who needs her care. I hardly need say
that I am very sorry to part with her. She is as
charming and as affectionate to us as ever.

———

Weather very dry during the greatest part of
June, though with great variations of temperature.
In Suffolk, indeed, it seems to have been more
steadily and constantly dry than about London.

We had a very pleasant visit yesterday to
Hampton Court, to our dear Lady Muriel and
Courtenay Boyle, who are staying there with his
mother, Mrs. Cavendish Boyle. We went by road
(by Kew, Richmond, Twickenham and through
Bushy Park) : the day was beautiful, warm, but not
too hot. We were received very kindly and
pleasantly by Mrs. Boyle, most cordially by our
dear friends, and spent a most pleasant evening,
lounging in the beautiful gardens till half-past seven,
when we dined with Mrs. Boyle and her son and
daughter-in-law, and Sir Francis Boileau who was
another guest of theirs. Altogether it was a delightful visit : the only drawback being, that dear Lady
Muriel looks as if in delicate health. I fear she has
suffered in health from her great sorrow in the death
of her mother. I had not seen Hampton Court
for many years ; we used to be very familiar with it
when Mr. and Mrs. Horner and their family lived
at Hampton Wick, from 1847 to 1853. I was
much struck now with the majestic and venerable
air of the building (much more truly *palatial* than
either St. James's or Buckingham Palace), and with
the quaint beauty of the gardens.

The avenues of horse-chesnuts and limes are very
grand.

We saw the famous vine, which was planted in
1768, and which is indeed wonderful, both for the
extent of its branches and for the tree-like bulk
and massiveness of its trunk. It still bears
abundant bunches of fruit. In the garden, observed

1881. two very fine bushes of Mastic, Pistacia Lentiscus,
which I have very seldom seen in English gardens.
They are planted in large tubs, like the Orange
trees, and no doubt housed in winter like them.
From the size of the stems I should judge them to
be very old.

In the gardens:—Gum Cistus, several fine bushes
in abundant blossom and in fine condition, not
apparently at all hurt by the winter.

As we were standing in the garden we saw a pair
of Herons flying over our heads, with their broad
flapping wings.

July 4th.

The news of the horrible assassination of the
President of the United States : horrible, and most
strange. How common such horrors have become
in our time ! Who can believe that mankind is
improving ?

We heard of the death of our old friend Lady
Napier, the widow of dear Sir George. She has
been, for many years past, so nearly dead—being
in a state of imbecility,—dead in effect to us, and to
almost all her friends, that the separation of the
soul from the body scarcely makes any difference
except to those who were immediately about her :
and to herself it can only be a release. Long ago,
in the lifetime of her husband, and indeed later, she
was a very handsome and very charming woman,
not very clever, but with peculiarly winning manners
and delightful warmth of heart ; always a most kind

and cordial friend to Fanny and me. She was truly
worthy to be the wife of that most loveable and
delightful man, Sir George Napier.

July 6th.

After a long spell of fine warm weather, the heat
yesterday became tremendous : 92 deg. Fahrenheit
in the shade in London, and 94 deg. in Scott's
garden at Barton, by his letter. A most violent
thunderstorm came on towards midnight, and lasted
several hours, with very heavy rain ; and to-day, the
weather is very chilly and unsettled.

July 8th.

My brother Edward's 70th birthday. Thank
God, he appears to be in good and comfortable
health, and I hope and trust he may continue so.

Fanny not well, having caught cold. I drove out
with Minnie : called by appointment on Mr.
Sanford, who spent an hour in showing us his very
fine microscope.

July 9th.

Went to the new Natural History Museum
(Cromwell Road), and spent above an hour with
great satisfaction in the Geological and Palæon-
tological Galleries. These appear to be the only
departments as yet so far arranged as to be open to
the public. The main entrance, or gateway, superb;
the great hall, very grand and noble, with abundance of light. The great gallery of Palæontology

1881. strikes me as everything that can be desired:—
plenty of room, air and light; gigantic specimens
(complete or nearly complete skeletons) standing
singly along the middle space, and more fragment-
ary, very well labelled, in glass cases at the sides.
Mastodon Ohioticus, a complete skeleton, very
grand.   Elephas primigenius, a grand skull with
enormous tusks, from Ilford in Essex.*   Elephas
Ganesa, one of the species from the Sewâlik Hills
in India, with still more gigantic tusks.   A great
collection of other species of Elephant and
Mastodon, collected by Falconer and Cautley from
the Sewâlik Hills.   The Deinotherium, a fine
skull, complete except the lower jaw, which is
replaced by a plaster cast from the original in the
Darmstadt Museum.

The collection of teeth and bones of the Elephas
primigenius (Mammoth) is prodigious, from almost
all parts of Britain, and from Northern Europe and
Asia.   Also abundant specimens of Elephas
meridionalis, and other species from Southern
Europe: and the curious Pigmy Elephant, found in
the caves of Malta, seemingly not bigger than a
Shetland pony.

The great Irish Deer, Megaceros Hibernicus,
two very fine complete skeletons, male and female.

The Megatherium, a skeleton complete, though
made up from two individual animals—a monstrous
creature, striking from the very small skull and
brain in proportion to the huge body and limbs.

The Glyptodon—the great extinct Armadillo of

* This and the Ganesa in the middle space.

La Plata—a very fine specimen as well as many 1881.
fragments.

Deinornis—complete skeletons of two species—
one larger than the Ostrich.

Eggs of Deinornis.

One sees very easily how different the extinct
Irish Deer is from the Elk, in shape and proportions, especially in the length and uprightness of
the neck and the proportionally small head. It is
truly a gigantic Fallow Deer.

Æpyornis, the gigantic bird of Madagascar, of
which the eggs only have been found. Two of
these eggs preserved here, each large enough to hold
two gallons.

July 11th.

Weather brilliant and excessively hot.

Very glad to find that all the accounts of the
great Review at Windsor on the 9th are so favourable, and that it is generally admitted to have been
a splendid success, almost without a drawback.
This is really surprising as well as agreeable.

We went (Minnie with us) into the National
Gallery to see the new Leonardo da Vinci. I
remember having seen it many years ago, exhibited
in the British Institution, and being struck by its
singular colouring. It is certainly a remarkable
picture, and I should have thought it (as to the
faces at least) a very characteristic specimen of
Leonardo; but I am told that its claim is disputed
by some critics. The scenery is very singular;

1881. the figures are placed in what appears to be a stalactite cavern: through the opening of this is perceived the bright blue sea, and a strange blue light appears to be reflected from the sea into the cave, and to give a rather ghastly and almost unnatural tint to all the figures. It reminds one strongly of description of the *Grotta Azzurra* in Capri.

Afterwards I spent an hour by myself very pleasantly in the Jermyn Street Museum, looking at the fine and well-arranged collection of beautiful minerals.

July 12th.

We went first to Cavendish Square, picked up pretty Blanch Sanford, took her with us to the Zoological Gardens, where we joined Lady Tollemache, and spent nearly two hours very much to my satisfaction in looking at the animals. The weather splendid, and most of the creatures in high glee, especially the maccaws and cockatoos. The greatest curiosity, a Koala, or native bear of Australia, the first, I believe, that has been seen alive in Europe. It is a young animal, and (being nocturnal in its habits) appeared very sleepy and lethargic, but Mr. Bartlett says it is lively enough at night. As we saw it, it perhaps suggested rather the idea of a sloth than anything else, though it was not hanging by its claws, but curled up in the fork of a branch. Its fur very long, soft, loose and fluffy, of an ashy-grey colour; face (in front view) broadly triangular;

ears, very large, or appearing so, from the quantity 1881.
of thick hairs with which they are covered. I could
not see its eyes or mouth. Claws of fore-feet very
large and strong and much hooked. It is a vegetable
feeder, and does not seem to eat anything willingly,
except leaves and shoots of eucalyptus.

Another novelty, the *Insectarium*, a room fitted up
for the rearing of butterflies, Moths, and other fine
insects, from the larvæ. Here are a few superb
moths, already reared and alive, in particular that
of which the larvæ feeds on the Ailantus: also living
caterpillars of Papilio Machaon, with their proper
food ; and a great many chrysalises.

In the Fish House :—Two species of those very
curious birds, the darters, the one Indian, the other
South-American. Their wonderfully flexible and
slender, snake-like neck, small head and slender,
sharp beak. But we did not this time see them in
the water, as I did once, a few years ago : they then
look more than ever strange and *un-bird-like.*

Toucans—several species in the Parrot House,
remarkable for the beautiful, bright colours of their
beaks as well as of their plumage.

Wapiti Deer (Cervus Canadensis) grand animals ;
the horns of the male really marvellous for their
size and intricate ramification—look as if they must
be very cumbrous and burdensome to the animal
itself.

———

July 13th.

The weather, splendid, as it has been ever since

1881. the 9th.   We (Emmie with us) went to Lady
Holland's garden party at Holland House.

_____

July 14th.

We visited the MacMurdo's at Rose Bank.

_____

July 15th.

We spent about an hour in the Royal Academy
Exhibition, and saw it comfortably, for (as we began
soon after ten) there was no crowd.   As much
delighted as before with *Long's* " Diana or Christ."
It is a noble picture, and one which deserves to
make a lasting impression.   Otherwise I think the
best pictures here are landscapes and sea pieces
(or water pieces) and some animals.

" Wargrave," on the Thames, I suppose, by *Vicat
Cole.* An exquisite, calm, still, peaceful, river scene.

" August Days," by the same, in much the same
style, and charming.

" Diamond Merchants, Cornwall."   *J. C. Hook.*
A beautiful bit of rocky sea-coast ; the colour and
transparency of the water quite exquisite.

" Mountain Tops."   *J. Mac Whirter.*   A very
striking scene, and forcibly reminding me of what I
saw many years ago from the top of Snowdon :—the
craggy summits appearing partially amidst the
rolling mists, and a clear background of sea and
distant hills showing beyond.

" Streatley."   *Vicat Cole.*   One of his delightful
Thames views, very like what we saw last year
when with the Charles Hoares, near Pangbourne.

" Mid-Channel."   *H. Moore.*   A very peculiar   1881.
and remarkable sea-view—most strictly so—nothing
but sea and sky ; the water wonderfully painted.

———

July 19th.

Dean Stanley died last night ; a very great loss
and heavy misfortune to the public as well as to his
private friends. I cannot say that I was ever intimately
acquainted with him, or qualified to judge properly
of what he was in private life ; but it is evident that
he was loved as well as honoured by all who really
knew him.   As a leader of religious thought,
and a writer on religious subjects, his death is a
national and irreparable loss ; there is not likely to
be any one in our time who can at all replace him.
As a champion of liberal and enlightened views of
religion, he was, I suppose, superior to anyone else
in our time : less stern than Arnold, free from the
occasional eccentricities of Kingsley, more intelligible than Maurice.   His books—The Memorials
of Canterbury and of Westminster, the Sinai and
Palestine, the Lectures on the Jewish Church, are
quite delightful.

The obituary article on Stanley in *The Times* of
this morning, is excellent.

———

July 22nd.

Rose Kingsley, to whom I wrote the first news
that the Dean's state was hopeless, writes thus of
him.   " I can scarcely believe it yet.   To the

1881. " Church, to the nation, to his innumerable friends,
" his death seems one of the greatest of possible
" losses, and when I think of that delightful
" afternoon at Kew, of our walk to the ' Lotus ' (to
" the house for tropical aquatics), where the Nelum-
" bium was in flower; of the extraordinary spirits
" and life which he showed then, it is quite
" impossible to realize that I shall hear the tones
" of that voice no more."

———

July 26th.

We returned home to Barton on the 22nd,
but Fanny has ever since been so ill as to make me
very unhappy.   Her illness indeed began in London
in the night of the 19th, and was certainly owing
to the overpowering heat; but on the 22nd she
seemed so far improved that Mr. Cameron
sanctioned her coming down hither; it was a sad
mistake however, for she was again very ill in the
night.

———

August 5th.

My journal has fallen into arrear.   I must pick it
up where I broke off on the 22nd of last month.

———

August 7th.

It is a great mercy that we have been blessed with
a week of such splendid weather for the beginning
of the harvest.   It must have made great progress
during six or seven days of almost unclouded sun-

shine, and we may hope that a great deal, at any rate, of it will be got up in fine condition. The prospects of the farmer certainly look more hopeful, or less dismal, just now than they have done for a long time past.

The 4th and 5th were really hot days, though not equal to some of those in the latter part of July.

A dismal and melancholy change in the weather since I wrote the last. The rain began on Monday the 8th, and ever since has fallen heavily during part at least of every day, throwing a dismal shade over the hopes of the harvest, which were before so bright. On Thursday the 11th, the rain was excessively heavy; yesterday it poured the whole day, and to-day has been little better. Unless it please God to give us fair weather, the harvest is ruined,— and what will become of the farmers, or the landlords either, I really do not know.

Arthur arrived at 11 o'clock last night, quite unexpectedly, having come in thirty-seven hours from San Sebastian without stopping, unless for meals. He is safe and well, but very thin, and (as might be expected), much sunburnt. I forget whether I noticed in my journal his setting out for Spain. He has been travelling in the North of Spain (from Irun as far westward as Oviedo and Gijon), in company with Clements Markham, who

1881. most good-naturedly offered to take him as his companion and assistant in a tour which he (Markham),
was going to make in those provinces. I have not
yet had time (as we have company in the house),
for enough conversation with Arthur to judge of
how far he has profited by his expedition, but he
seems to have greatly enjoyed it. He has brought
me some small specimens of iron ore from the
famous mines near Bilbao.

I see in the newspapers that the Irish Land Bill
has received the royal assent, and so is now part
of the law of the land—till it is overturned.

I suppose, since 1832, no bill has been so long
and obstinately and wearisomely debated. I dislike
it much ; it is a long step towards Socialism, and it
seems admitted to be contrary to all the rules of
political economy ; but, in the state which things
had reached—through the weakness of the ministry
and the outrageous violence and wickedness of the
Irish faction. I suppose the allowing the bill to
pass was the lesser of the two evils. At any rate, it
is fortunate that a serious quarrel between the two
Houses was escaped. Now, will this new law have
any effect at all in quieting or reconciling Ireland ?
I doubt it exceedingly.

Archdeacon and Mrs. and Miss Chapman arrived.
Mr. Beckford Bevan and his daughter and son, Mr.
and Miss Lushington, Mr. and Mrs. and Miss
James dined with us. Arthur gave me some
specimens of iron-ore from Bilbao.

———

Isabel (Hervey) Locke and her husband arrived. A large party from Ampton and Stowlangtoft came to luncheon:—Lady Rayleigh, Lady and Miss Frere, Agnes Wilson, Mrs. Arthur Wilson. Mrs. Storrs and Captain Harris dined with us—Mrs. Storrs very pleasant.

Thursday, 25th.

The Chapmans went away. Lady Molesworth, Dorothy Hoste and the Miss Phippses came to luncheon. Mr. Abraham and his daughter Louisa dined with us.

Sunday, 28th.

A beautiful day. We went to morning Church with Mr. and Mrs. Locke and Arthur, and received the Communion. Read some of Paul's Letters to his kinsfolks.

Wednesday, 31st.

A long and fatiguing discussion between us two, Scott and Betts, as to Gittus's farm.

A pleasant visit from John Herbert; we walked a little way with him, and were caught in the rain.

Fanny read to me Arthur's journal in Italy—very pleasantly written.

Thursday, 1st September.

Dear Arthur set off for Sandhurst in very good spirits.

1881.   Read part of Augustus Hare's article on Arthur
Stanley.

Wrote on business to Mr. Baldwin.

LETTER.

Barton, Bury St. Edmund's,
September 1st, 1881.

My Dear Katharine,

I believe you have experienced the same
sort of dreadful weather in Scotland that we have
had here, so we can only condole with each other,
and indeed with all our friends who have anything
to do with land, for the deluge seems to have been
almost universal.   Scott and Mr. Greene say that
it is the worst harvest they remember in their time ;
indeed it seems a hopeless business : and the
probable consequences are very serious to contemplate :—not only a great falling off in our (the
landlords) rents, but what is worse, a great number
of labourers thrown out of work through the inability
of the ruined farmers to employ them : — con-
sequently, discontent and combinations among the
labourers, encouragement given to Socialist agitators,
and danger to property.   I much fear that the
coming winter will be a very uncomfortable one.
Well, one can only say that it is God's will, and
therefore it must be right; and this is better than
the Mahommedan conclusion, that it is *fate*.

(September 4th). — We have both of us been
delighted with the article on Arthur Stanley, in
Macmillan, by Augustus Hare ; it is charming,

especially the anecdotes of his childhood and
youth. Fanny has also very lately read to me
Stanley's Sermons on the Beatitudes, which are
beautiful. What an admirable man he was, and
how generally he is valued now that he is gone! I
wish I had known him better.

I have begun to read Sir John Lubbock's Address
to the British Association at York, which promises
to be very instructive. I am interested in the
accounts — in your's and other's letters — of
Kinnordy, how changed it must be! I can hardly
(indeed not at all) imagine it in its new state.
Are the gardens and grounds altered in at all a
corresponding way? I hope not. I have such a
delightful recollection of it in '44. This dismal
season has been by no means so bad for the garden
and the arboretum as for the farm, indeed very
favourable for grass and trees; of the latter a few
are just beginning to show beautiful touches of
yellow and crimson here and there: particularly
the Liquidamber, which is full in view of my
dressing room window, and one or two of the Red
Horse Chesnuts. Our Agapanthuses are flowering
very well this year, and we have a Magnolia in
blossom close to the library window. In the houses
I do not think there is anything new except a
beautiful scarlet Clerodendron: but the Allamanda
is very fine.

I presume that Fanny has told you all about
Arthur and ourselves. We are both of us, I am
thankful to say, tolerably well in health, but the
wet dull weather does not agree with either of us,

1881. and is depressing to the spirits : and when there is this physical cause of depression, it is more difficult to bear up against vexations affecting our worldly concerns.

We are looking forward with pleasure to Susan's arrival. I hope she will not find Mildenhall very dreary.

Ever your loving brother,

CHARLES J. F. BUNBURY.

JOURNAL.

Friday, 2nd.

Rain, rain, rain !

Received a pleasant letter from Edward from Oban.

Read remainder of Hare's very interesting article on Stanley.

Saturday, 3rd.

A gleam of sunshine.

Mrs. Storrs came to luncheon—very pleasant.

Began to read Sir John Lubbock's address to the British Association at York.

Monday, 5th.

Had a good drive with Fanny in the open carriage.

Mrs. Paley and Mr. and Mrs. Burrows came to tea.

Tuesday, 6th.

Had a pleasant drive with Fanny in the open carriage. Business with Scott. Visit from Admiral and Mrs. Horton.

———

Wednesday, 7th.

Drove with Fanny. Saw Lady Hoste and Lady Molesworth. Mr. Sinclair and his sister came to luncheon and spent the afternoon with us; very pleasing, cultivated young people, and *she* very pretty.

———

Friday 9th.

Drove with Fanny; We visited Lady Rayleigh at Ampton, and Mrs. Horton at Livermere.

———

Saturday, 10th.

Mrs. Martineau and Mrs. Wilson and Agnes came to luncheon, and were very pleasant. Mrs. Martineau particularly agreeable.

———

Wednesday, 14th.

A beautiful day. Dear Susan arrived, safe and well, after her long journey (without stopping) from Forfarshire.

———

Thursday 15th.

Began to read Taine's Ancien Régime.

═══

LETTER.

My Dear Katharine,

1881.   I return your M.S.S.* which both Fanny
and I have read through, and of which she will,
no doubt, write her opinion. I think them all good,
and can see no reasonable objection to the printing
them as an appendix to the Correspondence. The
first of them (from *The Academy*) and the last (from
The Echo) appear to me the best of them; the
second also very good, especially as pointing out
the very remarkable proof which he gave, of a
truly philosophical spirit in adopting a whole new
theory late in his life, and shaking off that which he
had so vigorously and eloquently supported.

Perhaps I might wish that some notices from
personal friends might be inserted, as was done in
the instances of Mackintosh and your uncle and of
Kingsley, but I know the objections to such a
proceeding. I wish some good writer could describe
Lyell as I saw him in Madeira, standing on the
edge of an enormous precipice, with the sheer
descent of many hundred feet behind him, perfectly
thoughtless of danger, while he eloquently expounded the structure of the opposite cliff.

I am very glad you are so near the end of
your work, which I am sure will be very interesting; and I wish you well through the index, which
I take it, will be the most tiresome work of all.

* Charles Lyell's Life, which Mrs. Lyell was writing.

Dear Susan arrived safe and sound, and seemingly 1881. less tired than I should have expected after so long a journey; she is looking well, and I hardly need say that she is very agreeable. She gives very good accounts of the families at Shielhill, &c.; of course you have heard of us from her.

We have now, at last, beautiful weather, but too late, I fear, to save the harvest. The crop of *haws* is prodigious, and so is that of the fruits of trees and shrubs in general, both native and exotic; Euonymus latifolius particularly beautiful. There is a promise too of fine autumnal colouring. Liquidambar already in beauty. No time for more at present.

Ever your loving brother,
CHARLES J. F. BUNBURY.

JOURNAL.

Tuesday 20th

The sad news of the death of President Garfield.

Saturday, 24th.

Read Hooker's Address to the Geographical Section of British Association.

Drove with Fanny and Susan; we paid a long visit to Miss Waddington's, meeting there Mrs. Thornhill and Mr. Image. Edward arrived.

Monday, 26th.

Had a very pleasant drive with Fanny and Susan through lanes between Rougham and Bury.

Tuesday, 27th.

Edward went away. We (Susan with us) drove
to Mildenhall in the afternoon, had a pleasant drive
and found everything in perfect order,—thanks to
Agnes Fincham.

————

Wednesday, 28th.

Mildenhall. A beautiful day. We visited the
boys' school: then drove with Susan to the nursery,
saw Julia Betts and her children: then (with G.
Betts accompanying us on horseback) visited the
new plantation on the Hill: out again after
luncheon, visited new plantation on Holywell Row
road. Mr. and Mrs. Livingstone came to tea.

————

Thursday, 29th.

Mildenhall. Another beautiful day. A visit
from Mr. Isaacson: saw also the Livingstones.
Miss Bucke came to luncheon. We returned home.

————

Sunday, October 1st.

A most beautiful day. We went with Scott
through several of the plantations and shrubberies,
marking trees, &c. Dear Minnie and her brother
arrived. Mr. Harry Jones dined with us.

————

Monday, 3rd

Cloudy and cold. We received news of the
death of poor Phillip Miles at Clifton. Dear Susan
Horner went away. We have been driving with her

every day, and her society has been a great pleasure
to us. Louis and Fanny Mallet and the Livingstones arrived. Mr. Lushington dined with us.

Tuesday, 4th.

William Napier, the Yeatmans, the Goodlakes,
Admiral Spencer, two Miss Thornhills arrived.
The Livingstones went away. The John Paleys
dined with us.

Wednesday, 5th.

Dear Joanna arrived. Lady Hoste, Dorothy,
Mr. Greene, Sir W. Hoste, Mr. Bevan and his
daughter and Lothair de Bunsen dined with us.

Thursday, 6th.

Lady Gage and the Victor Paleys dined with us.

Friday, 7th.

John Herbert went away.—I very sorry. Walked
with the ladies. Clement arrived. Mr. and Mrs.
Bland and the Miss Thornhills dined with us.

Saturday, 8th.

Dear William Napier went away ; the Yeatmans
went early.

Monday, 10th.

Louis and Fanny Mallet went away. Had a
pleasant walk with Joanna.

1881.   Tuesday, 11th.

The dear little Seymour children, Charlie and
Willie, arrived late.   Admiral Spencer and Harry
went away.

———

Wednesday, 12th.

Had a very pleasant walk with Joanna, Minnie
and Mrs. Goodlake.

———

Thursday, 13th.

Scott gave me the produce of the allotment Rent
Audit.   John Herbert came to luncheon from
Ampton—also Mr. and Mrs. Barber, who stayed
long.   We planted a tree for Joanna.

———

Friday, 14th.

A furious gale of wind all day—I could not go
out.   Dear Joanna went away—I was very sorry to
part with her.   Wrote to Edward, to Biarritz.
Miss Yeatman arrived.

———

Saturday, 15th.

The Goodlakes went away after luncheon.

———

Tuesday, 18th.

The dear Arthur Herveys, the Livingstones and
other guests arrived, but I was confined to my room
with a cold, where I had a pleasant visit from the
Bishop.

———

Pleasant visitors. Had a post card from Rose Kingsley with accounts of her mother's dangerous illness. Had very pleasant talk with Lady Arthur and Mrs. Wilson.

Confined to my room till even. Most of our guests went away. A very pleasant dinner and evening with the Arthur Herveys and their son John, Minnie and Harry.

The dear Herveys went away. Clement arrived.

My Barton rent day ; but I was not able to do more than to go down for a few minutes to greet my tenants at their dinner.

Dear Kate and Mr. Hoare, Sir Robert and Lady Cunliffe, Miss Bevan and Mr. Rodwell arrived. A pleasant evening.

Went out in the landau with Fanny, Kate Hoare and little Charlie. We drove to Fornham St. Martin and back another way. Sir Lambton and Lady Loraine, Mrs. Wilson and two of her daughters came to dinner. The Clements Markhams arrived.

Weduesday, 26th.

The dear little Seymour children went away—also Mr. Rodwell.   Drove in the landau with Fanny, —very cold and stormy.

Spent much time in reading Mr. Markham's journal of his tour in N. of Spain, which he had lent us.   Mrs. Wilson and two of her daughters came to stay.

Thursday, 27th.

Montagu MacMurdo went away early.   Drove with Fanny in the landau.   Went on reading Clements Markham's Spanish Journal.

Friday 28th.

Gave my Cape book to Sir R. Cunliffe.   The Cunliffe's, Markhams and Wilsons went away.   My cold not yet gone.

Monday 31st.

Thermometer down to 26 deg.   Day fine, bright, and sunny and still, but very cold.   Dear Cissy and Emmie arrived.

[*November 1st.*   During the last few weeks, Sir Charles was reading Taine's " Ancien Regime," " The Odyssey," books 3 and 4.   " Arnold's Lectures," and the 3rd vol. of his " History of Rome," Donne's " Tacitus," and Wallace's " Island Life."   He also went on writing his " Botanical Fragments.—*F. J. Bunbury*].

The Barnardistons (4), Mr. Sanford and his daughters, Nellie and Blanche and Reginald Talbot and Cecil arrived. Read most of the November number of *Proceedings of the Geographical Society.*

Wednesday, 2nd.

All the party except Cissy and Emmie, Mr. Sanford and myself, went to the Bury ball.

Emily and Caroline MacMurdo had arrived in afternoon.

Thursday, 3rd.

Our dance—very much liked.

Friday, 4th

Walked through the garden and arboretum with Lady Florence. Most of our party went to a dance at Hardwick. Read some of " Tom Jones."

Saturday, 5th.

Had a pleasant walk with Lady Florence (who left us after luncheon), Minnie and Mr. Sanford.

The Barnardistons, the two MacMurdo girls, Cecil and three young men went away.

Monday, 7th.

The Sanfords went away—I like them very much.

1881.   Tuesday 8th.

Dear Cissy, Emmie and Harry went away ; Minnie alone remaining with us. Mr. Lushington and Scott came to drink tea with us, and to discuss on evening school.

———

Wednesday. 9th.

Had a pleasant walk with Minnie. Began to read (second time) Lecky's "England in Eighteenth Century."

———

Friday, 11th.

A visit from the Duke of Grafton.

———

Sunday, 13th.

A beautiful day, almost summer-like. We went to morning Church, Minnie with us.

———

Monday, 14th.

Dear Minnie went away to Ampton. Went on reading Lecky.

———

Wednesday, 16th.

Dear Minnie came back.

———

Sunday 20th.

Again a beautiful day. Read prayers with Fanny.

Walked in the garden after luncheon with Fanny, 1881.
Sarah, the two dear little boys Minnie and Lady
Grey.

—— ——

Received the " Life of Charles Lyell," and read two
chapters of it.   Wrote to Katharine.

═══ ═══

LETTER.

Barton, Bury St. Edmnnd's.
November 21st, '81.

My Dear Katharine,

At last *the* book has arrived, and I return
you a thousand thanks for it.   It looks delightful,
and I expect a great deal of enjoyment in reading it.
I have already read the first two chapters.   I am
sorry there is no more of the Autobiography, but it
is curious how seldom people seem able to persevere
long in that sort of work.   His history of his school
days is very entertaining, but it makes the same
impression on me as all other histories of corresponding experiences have made—it makes me feel
thankful that I never underwent the like, and never
shall, at least in *this* life.   I admire and honour you
for your perseverance and the good work you have
done in this book, and I am heartily glad to hear
from Edward that it has had such success at
Murray's sale.

The men here are still employed in clearing away

1881. the ruins produced by the storm on the 14th of October; I do not remember such havoc here—so many noble, picturesque, old oaks in the park shattered to pieces, or so mutilated as to be mere unsightly stumps. It is true that most or perhaps all of them were more or less decayed, but they were not the less picturesque. In the arboretum and immediately about the house, we have suffered much less; the principal losses have been a beautiful bird cherry tree, and the best of our two Judas trees, the other having been destroyed in the January storm. On the whole our losses in the arboretum have been remarkably slight, considering; and the two great oaks opposite to the S.E. front of the house have scarcely lost a twig.

By the way (though you will not easily perceive the connection of ideas), to go back to your book—I am delighted with the view of Kinnordy; it is so interesting to me to have such a record of a place so interesting to me of which I have such a delightful memory from our first visit, and such a painful one from the second. And it is so much the more interesting as the house is undergoing such a metamorphosis.

We have in the house at present, Sarah and Albert, and their two dear little boys, Minnie, Lady Grey, and Edward—all very pleasant.

I will write to you again when I have made some more progress with the book.

With best love to Rosamond, believe me,

Ever your loving brother,

CHARLES J. F. BUNBURY.

JOURNAL.

A fine day.   Went on reading " Life of Lyell."
Mr. and Mrs. Drummond, Lord John and Rev.
John Hervey arrived.   Lady Hoste and Mr. Greene,
Mr. and Miss Lushington dined with us.

Wednesday, 23rd.

A brilliant day.   Had a pleasant walk with Lady
Grey and some of the other ladies.

Mr. and Mrs. Sancroft Holmes arrived, also
Harry.   Mrs. Wilson, Agnes, and Mrs. John Paley
dined with us.   Went on with " Life of Charles
Lyell."

Thursday, 24th.

Mr. and Mrs. Bland, Miss Newton, Miss Waddington, and Mr. Guise dined with us.

Saturday, 26th.

Had a very pleasant walk with Mrs. Holmes.
Dear Sarah and Albert left us.   Lord and Lady
Bristol, Lady Mary and Mrs. and Miss Wilson
came to luncheon.

Monday, 28th.

Had a very pleasant walk with Lady Muriel Boyle (who had come on Saturday). Courtenay Boyle went away very early. Mr. and Mrs. Holmes, Colonel and Mrs. Corry, and Edward after breakfast. Mr. Zincke arrived.

—— ——

Tuesday, 29th.

Had an extremely pleasant walk with dear Lady Muriel; she went away in the afternoon. Clements Markham arrived; we all went in the evening to hear his lecture on the Basque Provinces—very interesting.

—— ——

Wednesday, 30th.

Fine, but cold. Lady Grey, Lord John Hervey, John Hervey, Mr. Zincke, and the Wilson girls went away.

—— ——

Thursday, 1st December.

Mr. Markham went away. Minnie, Harry, and the two dear children remaining.

—— ——

Saturday, 3rd.

Dear Minnie went away. I was very sorry to part with her; but dear Sarah arrived in the afternoon.

—— ——

Sunday, 4th.

We went to morning church with Sarah, Charlie, and Harry; we received the Sacrament.

—— ——

Monday, 5th.

Dear Sarah and her two lovely children went away in morning—Harry in afternoon.

LETTER.

My Dear Katharine,

I have now read through the first volume 1881.
of *the* Life, and have read it with very great
pleasure. The letters and journals are really
delightful, and I admire both your industry in
collecting and your judgment and taste in selecting
them. I think they will be very generally enjoyed,
there is so much liveliness and variety in them.
Those written during his travels in Auvergne, Sicily,
and Catalonia, are particularly entertaining,—so
full of eagerness, fire and impetuosity, and at the
same time full of matter. It is very remarkable
how early he fixed upon geology as the pursuit of
his life, and how steadily he kept to it : and also,
how early he espoused that essential part of the
Huttonian theory which ripened into *his* especial
doctrine. I like very much the letters written at
Kinnordy (and wish there were more of them) with
their mentions of the *Sky-Parlour*, the Clune hill,
the Cavity, and other points in which I always feel
an interest. I wish I could have a more distinct
idea of Bartley, but I suppose there is no print of
it, or you could not conveniently bring it in. One
of the many countries which I always regret not
having seen, is the New Forest. To be sure, it is
not very remote. Lyell's letters on strictly
geological subjects are sometimes hard to under-

1881. stand: and this I often remarked while he was among us; nothing could be more beautifully clear than his style and treatment of subjects when he wrote deliberately, and with a view to publication: but in his letters on the same subjects, he was often obscure from rapidity and conciseness, giving only rapid sketches or hints of his views. Altogether, as far as I have gone, I think your book gives an admirably true idea of him.

This day has been excessively wet and dark, but on the whole we have had in the last 3 or 4 weeks an uncommon allowance of fine and mild weather, and the number of plants in flower in the open air is quite remarkable. Of wild flowers which I have observed in flower, there are Daisy, Dandelion, Furze, Ranunculus bulbosus, the White and Red Dead-nettles, Shepherds' Purse, and Hieracium Pilosella. Of the garden plants I will send you a list when I have time; most of these are survivals (as all the wild plants I have mentioned are), but the Christmas Rose, Anemone coronaria, Jasminum nudiflorum, and Forget-me-not are *anticipating* their usual flowering season.

We are now actually alone, dear Sarah with her charming children having left us this morning, and Harry in the afternoon. We expect Arthur about the 15th, from Sandhurst, but I hope we shall be pretty quiet most part of the month, and have some undisturbed time for reading and writing (to say nothing of " casting accounts," which in these days of deficient rents, is not a very agreeable occupation).

I hope your accounts from Scotland are com- fortable, and that you and Rosamond are quite well.

Believe me ever,

Your loving brother,

CHARLES J. F. BUNBURY.

JOURNAL.

Thursday, 8th.

Had a pleasant drive with Fanny in the pony carriage. Received a welcome letter from Rose Kingsley, with a happy account of her mother.

Read the Acts 13, with the new commentary.

Tuesday, 13th.

Had a drive with Fanny in the pony carriage. Read part of chapter 15th of Acts with the new commentary and the Greek text. Answered a letter from Mr. Yeatman as to seconding him at the Athenæum.

Thursday, 15th.

Dear Arthur arrived from Sandhurst.

Friday. 16th.

Our dinner party. Mrs. Wilson and Agnes and another daughter, Lady Hoste, and Mr. Greene, Dorothy and Willie Hoste, Mr. and Mrs. John Paley, Mr. and Mrs. Victor Paley, Mr. and Mrs. Dunlap, Patrick Blake.

Sunday 18th.

Ground covered with snow in morning.

———

Monday, 19th.

Finished reading " Life of Charles Lyell."

———

Tuesday, 20th.

A long discussion with Fanny and Scott about rents.

———

[*December 20th.* The weather was most beautiful during all November and part of December. Sir Charles's studies during these weeks were the following :—

" Charles Lyell's Life," by Mrs. Lyell.
Sara Coleridge's Letters.
Taine's " Ancien Régime," and
The Acts, with the new commentary.
And he was also writing the volume he afterwards printed of " Botanical Fragments."—*F. J. Bunbury.*]

———

December 21st.

I must again, towards the close of another year, offer my most earnest and humble and heartfelt thanks to Almighty God for his merciful goodness and protection to me and to those whom I love. I feel indeed deeply that I have cause for much more gratitude than I can ever express. As I wrote last year, I can never be sufficiently thankful that 1881. my invaluable wife is preserved to me, and that we not only live together in unbroken harmony, but that advancing age seems only to bind us more and more closely together.

Just when we returned from London, towards the end of July, Fanny had an attack of illness which for a time alarmed me very much, as it seemed to resemble that from which she had been in so much danger when in Scotland in 1849; but by God's mercy it was soon subdued; and ever since she has been in her usual health.

As to myself, I am not strong, not capable of any great bodily exertion; but I am thankful to be able to say that I am, and have been all this year, very free from any painful or disabling or alarming illness, and considering my age, I may gratefully declare myself to be in good health. I have the comfort of almost always sleeping well, and I enjoy what I consider the greatest of *corporeal* blessings, unimpaired sight.

The year has been in one respect an unfortunate one; it has been marked by the deaths of many of our friends or relations. — Marianne Lyell; Lord Hanmer; Philip Miles; Hermann Pertz; Lady Napier: Edward Goodlake; Sir Edward Greathed; —all these have departed since last New Year's Day.

I hardly know whether to include James Spedding, for although I had known him a long time, we had never been intimate; yet I was sorry for his death. Two others, I must mention, because, though our acquaintance with them was but very recent, we had

1881. been very agreeably impressed by them, and they are public losses also ; Professor Rolleston of Oxford and Mr. Cox, the librarian of the Bodleian. A much greater man, and most generally and justly lamented must be added to the list, though we were not intimate with him, Arthur Stanley. The Church, especially the Liberal section of it, has certainly not for a long time suffered so great a loss as in his death, not in that of Kingsley perhaps, not even in that of Arnold. It is interesting to see how his merits have been acknowledged by tributes from the most various quarters, and it is evident that he was as much loved by those who knew him personally, as admired and honoured by those who did not.

The list of remarkable persons with whom we were not acquainted, who have died within the year, is large, and headed by two remarkable names ;—the Emperor of Russia, Alexander the Second, and the President of the United States. Then Lord Beaconsfield, Thomas Carlyle, Gould, the ornithologist, Burton, author of a valuable History of Scotland ; Count Arnim, Dr. Cumming, Trelawny, the associate of Byron and Shelley.

Another eminent man I must also mention, whom we much respected and liked, as much as our slight acquaintance allowed, Lord Hatherleigh. Sir Philip Egerton also must be added to the list.

We have lately had the agreeable news of three matrimonial engagements among our friends : Mimi MacMurdo engaged to Willy Bruce ; my nephew William to Miss Ramsay, niece of General Ramsay, and Ada Newton to Mr. Ridley.

The year 1881 has been remarkable for numerous 1881.
and terrible and destructive storms—destructive
both of property and life. The extraordinary snow
storm of January 18th, I have noted in my journal.
The effects of it are still visible, in the great gap in
the grove between the house and the east lodge. In
various parts of England it caused the loss of many
lives, but I heard of none in this neighbourhood.

The tempest of October 14th was terribly destructive to trees ; indeed I hardly think that within my
memory there has been such havoc made among
our old oaks, and elms both in the park and in the
pasture near the church ; and our neighbours
suffered as much as we. It was lamentable as well
as strange to see so many venerable oaks shattered,
split, mutilated, in some cases even twisted and
wrenched round. It is true that most or perhaps
all which thus suffered, were more or less decayed
but they were not the less picturesque.

I am thankful that the *arboretum* suffered no
important loss in either of these storms.

Certainly we have lost more trees in this one year
than in the whole of the previous years since 1860.
In the January storm the great destruction was
of firs, willows, and other tall and slightly rooted
trees, which were fairly torn out of the ground. In
October, the chief sufferers were oaks, and they
were not uprooted, but torn and shattered.

I will say little about public affairs, because I see
scarcely anything in them that is cheering or comfortable. In one respect only, I think, our position
is better than it was last year ; that we are not

1881. engaged in any war, great or small. The state of Ireland appears, as far as I can at all understand it, to be as bad as ever; and the contagion of Irish agitation, and of the encouragement afforded to it by Gladstone's Land Act, has led many discontented farmers in England to imitate in some degree their example. No doubt the English malcontents are as yet pursuing their objects in a milder way: they may probably have no intention of breaking the peace or breaking the law; but the schemes of Mr. Howard and the Farmers' Alliance clearly aim at altering the conditions of ownership of landed property. They are therefore decidedly Socialistic.

LETTERS.

Barton,
December 26th, 1881.

My Dear Susan,

I must write you a few lines to wish you many happy returns of your birthday and a happy. New Year. I trust that you and all your family party at Florence are enjoying good health, and will be able to welcome the new year with cheerful and comfortable feelings. As years roll on, and one begins to feel the infirmities of age, and still more when one reflects on the many who have gone before us, a "browner shade" (as Gibbon says) is

cast over the prospect: but we may be very 1881.
thankful when we have no more to complain of
than Fanny and I have.   The year which is now
nearly past, has taken away not a few of our friends,
though happily none of the nearest and dearest.

It is a great pleasure to have dear Katharine and
Rosamond with us now: Arthur Lyell left us
yesterday, and they seem well; when they leave us
we shall be for a little while quite alone, as our
Arthur is visiting other friends.

I have read the "Life of Charles Lyell" with
great delight, and am now going through it again,
not reading it straight through, but *picking out the
plums*, which are very many.   No doubt it is in
some measure incomplete; so—I take it—is every
biography (except, perhaps, Boswell's Johnson).   I
do think, however, that there are some little
omissions which might easily be corrected.   But I
think it admirably well done, and it has interested
me more than any other book I have read since the
"Life of Charles Kingsley."   Lord Campbell's
"Life" also was very pleasant reading, and gave me
a much higher opinion of him than I had before.   I
am now reading the first volume of Lecky's
"History of England," *for the second time*, for I
was quite vexed to find how much I had forgotten
it, though I read it less than four years ago; and
as I hope there will be two more volumes coming
out soon, I am preparing for them by refreshing
my memory of the former.   I find that my memory
is growing old; I have more difficulty than I used
to have in impressing durably on my mind what I

1881. read.—Not that I think this is *entirely* owing to the weakening of the memory by age; the greater multiplicity and variety of common things— especially of business, which one has to attend to as one grows older — accounts for part of the change.

Talking of growing old, Mr. Bentham, who is eighty-one or two, lately sent to the Linnæan Society an elaborate and masterly paper on one of the largest and most intricate families of plants. Pretty well for that age.

I have written no news to you; the reason is, that I know none. I know Fanny keeps you acquainted with the small events of our circle, and politics are too detestable for me to enter upon. Mimi MacMurdo's engagement to Willy Bruce, and Arthur's promotion to be a sub-officer at Sandhurst, have given us great pleasure.

I am not sure whether this will reach you in time for your birthday; but in any case I hope you will accept my hearty good wishes for the health and happiness of yourself and all your home circle for many years to come.

Ever your loving brother,

CHARLES J. F. BUNBURY.

Barton, Bury St. Edmund's,

December 30th, 1881.

My dear Edward,

I do not know exactly where you are at this present time, or whether you are in the way of hearing what has just lately happened at Bury— the sudden death of Lady Blake. It appears that on Christmas day, when she appeared to be in good health, she slipped in coming downstairs, and fell, but only on her elbows; she thought at first she was not hurt, and sat down to luncheon with her sisters, but before the luncheon was over, she felt uncomfortable, and went up to her room to rest; her sister following in a little while, found her lying on the bed in a state of insensibility, from which she never woke. I am very sorry. She was a very good, kind, amiable woman, and I believe will be lamented by many for her kindness of heart. Wherever you may be, I hope you are well, and that you will be careful of yourself. Both Fanny and I very heartily wish you a happy New Year, and as many of them as can be conducive to your happiness, for I do not think you—any more than we, would wish life to be prolonged after the power of enjoying it is gone. As yet, thank God, that is by no means the case; though I am by no means strong, and she often suffers from sleepless nights, we may both be very thankful to be as well as we are.

I hope you are enjoying your New Year in pleasant company.

With Fanny's love,

Believe me ever your affectionate brother,

CHARLES J. F. BUNBURY.

JOURNAL.

1881. The year is ending with a fine and mild day. The little yellow Aconite, Eranthis, is in full blossom in the shrubbery, and Snowdrops are beginning to peep above ground.

So ends 1881.

God grant that if I be spared to see the end of another year, I may be a wiser and a better man.

1882.

LETTER.

Barton Hall, Bury St. Edmund's,
February 4th, 1882.

My dear Katharine,

Very many thanks for your kind letter and good wishes on my birthday. It is a great mercy that at the age of 73 I find myself not only alive, but in such good health, and surrounded by so many blessings, for which I can never be thankful enough, especially for such a wife and such friends. I thank you very much for the gift of Stanley's "Church of Scotland," which I have no doubt I shall find interesting, at any rate more so than any one else could make the subject. I quite agree with you in your opinion of Stanley, both as a writer and a man. His death was a loss which can hardly be over-estimated.

Edward's article on Charles Lyell I think admirable in every way, and it is surprising how much 1882. matter he has compressed into the space without any painful excess of conciseness. It will bear reading over several times. That number of the *Quarterly* is a very brilliant one, I think, for the article on Count Montlosier is very entertaining,— that on the Jacobins (by Lady Eastlake, Edward tells me), very interesting, and that on Fishes very curious. The article in the *Edinburgh*, on the French Revolution, I *guess* to be by Mr. Reeve.

I am very glad you have got back to your dried plants, and have been studying them again a little.

This has been really a most beautiful day ; it is very seldom that the 4th of February is such a one. We have four species of crocus in blossom in the garden, two of snowdrop, and two of anemone; and to-day, for the first time this year, I saw the little celandine (Ranunc. ficaria).

It is time to leave off. Many thanks for your pretty card.

Fanny is tolerably well, I am thankful to say, and I very well.

Ever your truly loving brother,
CHARLES J. F. BUNBURY.

I am studying Bentham on Grasses; what a wonderful man he is, to do such work in his 82nd year. I am also studying Judd on Volcanos, but the mineralogical part of it puzzles me. That science has been completely transformed since my father and I studied it.

1882.   We have some very beautiful orchids now in
blossom in our hothouse.—

> Dendrobia nobile and Wardianum.
> Calanthe Veitchii and Phaius grandifolius.
> Odontoglossum Alexandræ is just past.

Barton, Bury St. Edmund's,
February 5th, '82.

My Dear Lady Louisa,

I thank you very much for your kind and
pleasant letter, which I received yesterday as you
intended. It was very kind of you to think of the
4th of February being my birthday, and I am much
fl atterd by it. Yes, I have actually completed my
73rd year. And I may well be very thankful to be
in such good health at this age, and still more that
Lady Bunbury is so. I am afraid that I can make
you but a very poor return for your pleasant letter,
as we have scarcely stirred from home since the end
of July, and nothing very remarkable has happened
either to us or to our immediate neighbours. We
have had from time to time several very pleasant
parties of friends staying with us, but it would not
be interesting to give you mere lists of names.

Besides estate and parish business, which in these
rent-reducing days is not particularly agreeable,
my chief occupation has been reading. One of the
most interesting books I have read is *Taine*
"L'Ancien Régime," a very full and copious, and
most interesting and remarkable picture of the
institutions and the condition of France, under

Louis the Fifteenth and Sixteenth; full of curious details, and most instructive as showing how things gradually ripened for the revolution.

Then we have been very much interested (as you may suppose) by the Life of our brother-in-law, Sir Charles Lyell, which has been written by Lady Bunbury's sister, Mrs. Lyell. It is in my opinion admirably well done, and extremely agreeable reading; there is a great variety of matter in the letters, so that those who do not care for geology will find much to entertain them.

I suppose you see English books at Rome, and especially the reviews and magazines. The January number of *The Quarterly* contains several good articles; an excellent review of the Life of Lyell which I have just been speaking of; a very interesting article on the French Revolution, by Lady Eastlake, I am told; an entertaining one on the Comte de Montlosier and a curious one on Earthworms. This last does not seem an attractive subject, but if Darwin is right, they are very wonderful animals.

I hope you and your Roman friends will have very fine weather for the Carnival, and that it will not disappoint expectation. Here the weather was astonishingly mild and fine all through last month, and even yet though the last few days have been colder there has been neither snow nor really hard frost. I wonder how long this will last. It is surprising what a number of flowers are already out in the open air—crocuses of four different sorts, snowdrops yellow aconites, anemones, and one or two others.

1882.   At Rome, I suppose the gardens are full of blossom.
—I am sorry the recent excavations have been so
unproductive.   We are very much pleased with
Mimi's engagement, though like you, we wish there
were more means.

Lady Bunbury desires her love to you.

Believe me ever yours affectionately,

CHARLES J. F. BUNBURY.

———

Barton, Bury St. Edmund's,
February 9th, '82.

My Dear Joanna,

I believe you will not yet have heard of
our sad loss.   Our dear, good, old kind Janet
Rennie died yesterday, about one o'clock in the
afternoon.   She had been ailing for some weeks,
but it was not till very lately that we were aware of
her having a dangerous illness, and the end came
very rapidly—almost suddenly.   Happily, she seems
to have had little or no suffering.   About ten or
twelve days ago, indeed, there seemed reason to
fear that she had a complaint in the throat which
might end in suffocation, and we were afraid her
end might come with much suffering; but about the
end of last week she seemed to get much better,
then on Monday there seemed to come a collapse,
a failure of the vital powers, and Dr. Macnab told
us there was no hope.   Yet on Tuesday evening,
when I saw her for the last time, she was up and
dressed, and received me with her old kind and

cordial manner, though she was much altered in 1882.
face, and could not speak so that I could understand her. About 11 that night, after many hours
of restlessness, she fell into a deep sleep from
which she never woke: so she passed quite tran-
quilly away.—It is very sad. There could not be a
more kind, faithful, devoted, loving creature, or one
more full of human kindness; generous and loving
to all, tender to animals as well as to human beings,
and with much refinement in her nature; earnestly
religious, too, I believe. It is a comfort to feel that
she had a peaceful and comfortable old age with us,
and her simple and loving nature must have made
her life a happy one. She was, I believe, between
71 and 72 years old, and next May it will be 32
years since she came to live with us. I shall miss
her exceedingly, especially when we come back from
any temporary absence, when the greeting of her
kind, cordial, smiling face, used to be one of my
great pleasures. I shall never find a friend more
devoted to us.

I owe you a great many thanks for your kind and
pleasant letter of the 2nd, and your good wishes on
my birthday. I have indeed great reason to be
thankful, that at the age of 73 I am in such good
health and enjoy so many blessings. My birthday
was a most beautiful day, one of the finest winter
days I ever remember; the weather now is colder,
and dull, but there has not yet been any snow nor
severe frost. The weather is very favourable for
farming operations, and I believe there are no
people out of work in the parish.

1882.   With best love and thanks to Susan and Leonora, believe me ever,

Your loving brother,

CHARLES J. F. BUNBURY.

Barton,
February 10th, '82.

My dear Edward,

The death which I now have to tell you of is one which causes us a great deal of sorrow. Our dear, good, kind old Mrs. Rennie died the day before yesterday, after a short illness: she had indeed been ailing for some weeks, but it was not till a few days ago that she was supposed to be in any danger; then there seemed to be a sudden collapse, and she sank very rapidly. Happily she suffered very little, and passed away in a profound sleep without a struggle. She is a very great loss to us. She had been with us nearly 32 years, and a more thoroughly kind-hearted, simple, loving, faithful creature, it would have been impossible to find. No one can supply her place to us. Fanny is very much afflicted, and I shall always miss her; and, as Arthur said the other day, all "the nephews" will mourn for her, she was so loving to them all. Indeed she was full of love and kindness to all, and to animals as well as to human beings. It is a great pang to part with such an old familiar friend; but so it is—

" Hæc data pœna din viventibus."

I thank you very much for your kind letter and

good wishes on my birthday, and was interested 1882.
by what you told me of your geological studies.
I hope you did not suffer from the fog on the 4th,
which the newspapers described as terrible in
London. Here, the 4th was a most beautiful day ;
one of the finest winter days I ever remember.
Fanny is, I am thankful to say, well on the whole,
though at present much depressed in spirits. We
are now quite alone, Arthur having gone back
to Sandhurst. With much love from Fanny, believe
me

Your affectionate brother,

CHARLES J. F. BUNBURY.

JOURNAL.

February 13th.

Our dear, good, kind old servant and friend, Janet
Rennie, died on the 8th, happily with little or no suffering, after a very short illness; she had lived with us very
nearly 32 years—ever since May, 1850; and she was
quite devoted to Fanny and me ; at the same time,
she dearly loved our nephews, especially Clement
and Arthur, and was warmly attached to Fanny's
family and to the Lyells; indeed her heart was
so full of love and kindness, that it overflowed on
all she lived with.

God grant that we may hereafter meet her again
in a better world.

LETTER.

My Dear Susan,

1882.   I have to thank you very much for your kind wishes on my birthday, and also for the photograph of the Correggio, which is so like ours. Fanny has written to you on this subject. I am much obliged to you for the information you have sent us about it, but 1 am quite content to let it remain in its previous state of uncertainty, as a doubtful Correggio, and an indubitably charming picture.

I thoroughly feel with you that "every birthday brings with it the recollection of accumulated blessings, however many the sorrows incident to this stage of our life." I feel indeed strongly how very much I have to be thankful for, how much more fortunate and happy I have been than I deserved. Yet as I grow older, and think of the many loved ones who have gone before me, I cannot help sometimes feeling a dread—not of death at all, but a dread lest many more of those whom I love should be called before me. Do you know what was thought by the ancient Romans (and very justly), to be the most awful curse on a bad man ?—

" Ultimus suorum moriatur !"

May he be the last to die of all his people. This is a lame and clumsy translation—no one English word can give at all the force of *suorum ;* it implies

kith and kin and friends. At the same time there is 1882.
a certain pleasure in thinking (when the loss is not
too recent) of the dear ones we have lost, and of the
happy days we have passed with them : to say
nothing of the hopes for the future.

I hope you liked Edward's *Quarterly* article on
Charles Lyell ; I thought it delightful, and there are
several other good articles in the same number.

I am now reading De Tocqueville's "Ancien
Régime," which is instructive, but very dry; a
great contrast to Taine in the manner of treating
the subject, though both agree very much in
their conclusions. I am going to read Count
Montlosier, which Fanny has found interesting.

I must, if I can find time, read Darwin's
" Earthworms," which seems to have excited an
extraordinary degree of curiosity ; and indeed he
makes them out to be very extraordinary creatures.

I hope this fine and mild winter agrees with you,
for I see by yours and Joanna's letters, as well
as by the newspapers that it is remarkably fine in
Italy as well as here. I like it very much and it
agrees very well with me, but certainly it does seem
extraordinary, and almost unnatural, after the
experience of several years past. I hear that many
people have bad colds, probably owing to im-
prudence. The mild winter is a blessing to the
poor, not only in the absence of suffering from severe
cold, but because field work has not been interrupted
by frost.

Our garden is beautiful now, with a profusion of
snowdrops, crocuses, anemones, primroses, forget-

1882. me-nots, and lilac scillas, and roses and many other
shrubs, are coming into leaf. Pray give my love to
all your family party. I am very much obliged to
Leonora for her letter and her good wishes.

Ever your loving brother,

CHARLES J. F. BUNBURY.

JOURNAL.

February 27th.

I see in the newspaper the death of Decaisne,
the greatest of remaining French botanists—a great
loss.

The weather very extraordinary all this month,
and indeed ever since the beginning of the year.
There has really been no winter as yet. No snow
at all in January; only one fall in February (the
15th) and then it melted very soon. No really
severe frost : the lowest the thermometer reached in
January, was 25 deg. Fahrenheit (7 deg. below
freezing point). There has also been an uncommon
proportion of fine and sunny days. The garden
beds are now beautifully gay with spring flowers,
and almost all are a full month earlier than last
year.

LETTER.

Barton,
March 8th, '82.

My Dear Edward,

Our friend Mr. Yeatman is a candidate for
the Athenæum, and writes to me that he hears the

ballot for him will come on in the week after Easter. 1882.
I had promised to second him (as his original
seconder is dead), if I should be in town at the
time, but this I cannot be. I should be very much
obliged to you if you would second him (unless you
are going out of town just then), and give him all
the support you can ; he is a particularly agreeable
man, and we are both very fond of him. I think
you must have met him either here or at our house
in London ; his wife, Lady Barbara, is one of the
Lady Legges, sister of Lady Louisa, Lady Octavia,
and the rest whom we know so well. His name is a
very singular one—*Huyshe Yeatman.*

Many thanks for your letter of the 5th, we were
both much gratified by the kind way in which you
speak of our dear Mrs. Rennie.

I am happy to say that Fanny is now quite well,
much better than she was a short time ago.

I was very glad to see in the papers that
Monckton Milnes had rallied from his attack, but I
am afraid it must leave its traces.

I have been much interested by a new novel
which no doubt you have seen—"John Inglesant;" I
think it remarkably well written, and very impressive
—also, in some respects, puzzling.

Our garden is full of the beauty of spring flowers :
most things quite a month earlier than last year.

Ever your affectionate brother,

CHARLES J. F. BUNBURY.

March 20th, 1882.

My Dear Edward,

Mr. Percy Smith, a short time ago, wrote
to tell me that he had definitively made up his mind
to resign the living of Barton, and would carry this
purpose into effect in the course of the summer ; he
has obtained the appointment of permanent
English Chaplain at Cannes, and he finds the
climate much better for both his wife and himself
than that of England. So I had to look out for a new
Vicar of Barton. I offered it to Mr. Harry Jones
who I found was about to resign his Rectory of
St. George's in the East, by the urgent advice of his
doctors ; and after taking some days to consider, he
has accepted it. Of course he cannot be, properly
speaking, appointed, till after Mr. Smith has
formally resigned. You no doubt know *about* Mr.
Harry Jones, though you may not know him personally : we do not know him intimately, but from all
that we do know, and from what we hear, I am
satisfied that he is a very good and very able man,
a cultivated man, and of great activity ; and I hope
and believe that he is of moderate views in church
matters. It is a very hazardous thing—the nomi-
nating a parson to one's own parish ; especially as
(like matrimony) it is irrevocable ; but I hope and
trust that my choice will turn out well for the parish
and for ourselves.

Many thanks for your last letter, and for your
consent to support our friend Mr. Yeatman. Fanny
and I have both had colds since I last wrote, hers
was for a few days rather severe, but I am thankful

to say we are both well again. The weather is 1882.
glorious.

We have had a very agreeable party in the house
—Aberdare and his daughters Lina and Sarah and
his son Willie, Lord and Lady Walsingham,
Lord and Lady Rayleigh, Lord John Hervey,
Courtenay Boyle and Lady Muriel, Professor
Hughes of Cambridge, young De Bunsen and
Clement, not to mention dear charming Mimi
MacMurdo, who is staying with us till her parents
return to England. Aberdare and Lord Walsingham
particularly agreeable,—but all were pleasant.

Most of them are now gone, but the Walronds
come to-morrow, and the Augustus Vernon
Harcourts on Thursday.

Where are you going for Easter?

I long for Lecky's two volumes.

Will you come to us next Monday, the 27th? or
would you prefer Passion week or Easter week?
We hope you will come one of these times.

With Fanny's love,
> Believe me ever,
>> Your affectionate brother,
>>> CHARLES J. F. BUNBURY.

JOURNAL.

March.

The atrocious attempt to shoot (or frighten) the
Queen, foolish and wicked as it was, has had one
good effect,—that of calling forth most satisfactory
demonstrations of loyalty and attachment to her

1882. from all quarters,—even from Ireland ; and indeed
I should think it may have done actual good in
rekindling and reviving those sentiments in some
cases where they may have been a little slumbering.

———

April 1st.

Exceedingly pleasant company in the house since
the 15th of last month.   On that day Courtenay
and Lady Muriel Boyle came to us, and remained
here till the 29th.   I need not say that her company was a pleasure to me ; she is a most delightful
being, one of the most loveable women I know, and
so warm-hearted, so kind, and shows such warm
friendship to us, that her society is a real delight.
She is very intelligent too, and without being
learned, has quite enough reading to keep her
conversation from frivolity.   Then Aberdare and
his two eldest daughters, Lina and Sarah Bruce,
and Lord John Hervey stayed with us from the
17th to the 20th, Lord and Lady Walsingham,
Lord and Lady Rayleigh and Professor Hughes
from the 18th to the 20th, the Theodore Walronds
from the 21st to the 25th, the Augustus Vernon
Harcourts from the 23rd to the 29th.   Aberdare is
a peculiarly agreeable man, with a really surprising
range of knowledge on a vast variety of subjects, a
great fluency of conversation, a great command and
happy choice of language, and very agreeable
manners.   He is also a very handsome man for his
age.   His daughters are very pleasant girls.   The
Walsinghams we had not seen since we visited them

at Merton in '78; Lord Walsingham is a very handsome man, and of very pleasant manners, very clear and well informed, an eager and accomplished naturalist, of extensive knowledge in more than one department of that science. Entomology is his special study, but he seems to have also much knowledge of birds, and at least enough of botany to be a very intelligent observer.

Augustus and Rachael Vernon Harcourt are quite as charming as I thought them on their former visit here, and when we were their guests at Oxford. The days when we were alone with them and the Boyles, were as pleasant as any in this social fortnight.

April 15th.

Dear Mimi (Emily) MacMurdo has been with us constantly ever since the 11th of March. It would be difficult to find a more charming girl, or a more agreeable companion ; though in some respects, one must say that she has not been highly educated : she has great natural abilities, a great love of knowledge and desire to cultivate her mind, a quick and lively intelligence, and, if I am not mistaken, much clearness of ideas. She appears extremely happy in her engagement with William Bruce, and they seem each truly worthy of the other.

William Bruce (Aberdare's 2nd son, and the eldest of my cousin Norah's children), is a really charming young man ; everything, seemingly, that can be desired either in intellect or morals. His mind is highly cultivated ; the extent and variety of

1882. his knowledge strikes me as quite remarkable at his age, and with this, he has a most pleasing modesty, which, perhaps (considering his age and all circumstances), is even more uncommon than his talents and knowledge. He really seems to be quite worthy of Mimi; they appear very much attached to each other, and I trust, please God, they have very bright prospects of happiness, though they will not be rich.

———

April 24th.

The news of Charles Darwin's death startled us on the 22nd. It is a great loss to science as well as to his many friends. Old as he was, (just eight days younger than I) he had so long been indefatigable in his pursuits, and had so lately given public evidence of his unabated mental activity and clearness of faculties, that one could not help expecting still more from him ; and at the first moment it seemed strange (though in reality perfectly natural) that such a fount of knowledge should be suddenly cut off.

He was decidedly the greatest naturalist of our time and country ; perhaps of our time without the limitation of country. What is most remarkable, he was not only transcendently great in the two departments (zoology and geology) to which he chiefly devoted himself, but he threw new and most important light on some branches of botany, particularly on the physiology of plants.

———

April 25th.   1882.

Joseph and Lady Hooker came to us on the 8th, and stayed till the 11th. They were very pleasant. He is looking fairly well, and seems to retain a good share of activity of body and mind ; talks not quite so much or so fast as I have formerly known him to do, but well and pleasantly, with great range and variety of knowledge.

LETTER.

Barton, Bury St. Edmnnd's.
April 26th, 1882.

My Dear Katharine,

I am sure you must have been very much grieved and shocked by the death of Charles Darwin.

Old as he was, and long as he had been a habitual invalid, he had so lately shown the un-abated activity of his mind, and clearness of his faculties, that one could hardly help expecting still more from him ; and it seemed almost strange that such a supply of knowledge should be suddenly stopped. It is certainly a grievous loss. He was decidedly the greatest naturalist of our time ; and however his theory of Natural Selection may be shaken or modified in time to come, his works are such storehouses of facts, carefully observed and well arranged and clearly stated, that they will be valued as long as natural history is studied at all.

1882.   Well I remember, the rainy day that I passed
with him in the old inn at Capel Cerrig in 1842;
and most agreeable he was; my real acquaintance
with him began that day, though we had been
introduced at the Geological a few months before.
South America was a subject in common which
quickly cemented our acquaintance.   I have very
seldom had an opportunity of meeting him of late
years, but whenever the chance has occurred I have
always found him very pleasant.   I am very glad he
was taken away by such a speedy death, and before
any decay of his powers.

Fanny's letters will have kept you well *up* to the
course of our quiet lives.   We had a pleasant visit
from the Hookers.   I am very glad you have had
comfortable accounts of your oriental travellers.

(April 27th).   Cissy and Emmie have arrived, and
also Major Godwin Austen, a son of our old friend,
who is quartered at Colchester.   We are going to
have a large dinner party.   The weather, after
having so long been delightful, has turned cold and
variable, and makes us feel rather uneasy as to the
prospects of the crops.   Another bad harvest would
be utter ruin both to tenants and landlords.   But
we must hope and trust as long as we can.

Barton is now in full beauty of spring, and I do
not willingly think of leaving it.   Everything is
astonishingly forward.   The lilacs are partly in
blossom, and our large horse-chestnut fully so; fully
a month earlier than last year.   I hope Frank is
much better than when we last heard from you

and that Leonard and Arthur will return to you in 1882. good health. We are both pretty well.

Ever your loving brother,

CHARLES J. F. BUNBURY.

JOURNAL.

April 28th.

My acquaintance with Charles Darwin began in the spring of 1842, at the Geological Society. A few months later in the summer of the same year, he and I happened to be staying at the same time in the old inn at Capel Cerrig ; and as a very wet day threw us much together, we had much talk, and our acquaintance ripened rapidly. South America in particular furnished us with abundance of interesting matter of conversation. I found Darwin extremely pleasant.

My opportunities of meeting Darwin afterwards were but few and rare, and for many years past especially rare ; but whenever the chance did occur, I always found him very friendly and pleasant. One of his peculiar characteristics, as it always struck me, was his unaffected modesty ; the entire absence of anything like affectation or pretension.

April 30th.

Yesterday was very stormy all day, and towards evening the gale became tremendous, continuing so till past midnight. The mischief has been great : —above all, one of our largest and finest Beeches

1882. (situate near the entrance to the arboretum), has been torn completely up by the roots. This was a beautiful and favourite tree, and is a great loss. It appears to have been perfectly sound, so that at first sight one wonders how it could have been so uprooted; but the disaster was probably owing to its situation—it stood in a sort of defile or pass, an opening between two groves, through which the south-west wind would be in a manner drawn in, and would blow with special fury.

Besides this great loss, the storm has been fatal to the last that remained of our great old white Willows, in the approach, from the east lodge: this also is torn up by the roots.

LETTER.

Barton, Bury St. Edmund's,
May 2nd, 1882.

My Dear Edward,

The storm on Saturday last did a great deal of mischief here; in particular, a very large and fine beech tree standing just at the entrance to the arboretum, was torn completely up by the roots. Another tree which has been uprooted was the last of our large old willows in the approach from the East Lodge.

I hope you have quite recovered from your cold.

Cissy and Emmy alone are with us now (Mimi having left us on Saturday), and I believe they will not stay beyond this week.

We talk of going to London about the 25th: I do 1882.
not look forward to it with much pleasure, but I
own it would not *do* to be always rusting and
rusticating.

I have heard the cuckoo to-day, for the first time
this year. The migratory birds seem late this
year, though the flowers and leaves are remarkably
early.

Poor Norman is very ill, but Dr. Macnab is
hopeful of his recovery.

With Fanny's love,

  Believe me ever,

   Your affectionate brother,

    CHARLES J. F. BUNBURY.

JOURNAL.

May 6th.

We received news this morning—in a letter to
Fanny from dear Lady Muriel Boyle—which has
agitated us a good deal with a mixture of joy and
sorrow. Courtenay Boyle is appointed Private
Secretary to Lord Spencer, and accompanies him to
Dublin—indeed I believe they have already started
—and Lady Muriel follows with Lady Spencer next
week. It is an honourable appointment for
Courtenay, and will have various advantages, so
that for their sakes I must rejoice at it. But as it
concerns *us*, I am very sorry; it is a great loss to
have two such friends—with whom we have now
become so intimate and to whom we are so much

1882. attached—removed to Ireland, so as for an indefinite time to be quite out of reach—out of all chance of meeting. I feel very much this separation from Lady Muriel whom I really *love*, and whose society I have so much enjoyed in the last few years. I hope I may live to meet her again, but of course, to a man of 73 years of age, any such prospect must be uncertain.

———

May 9th.

I had hardly finished writing the last page before we received the horrible news of the murder of Lord Frederick Cavendish and Mr. Burke in the Phœnix Park. A great many shocking and frightful crimes have been committed in my time, but I hardly think there has been any so atrocious as this,—so utterly unprovoked, for Lord Frederick had only just entered on his office, had had nothing previously to do with Ireland, and could not possibly have injured or annoyed any one there. Even the Irish newspapers have generally professed (how far honestly may well be questioned), their abhorrence of the act, and Parnell himself has written and spoken with very right and proper feeling on the subject. I have no doubt that this truly diabolical act has been committed by Fenians— Americanized Irish—some of O'Donovan Rossa's gang. The object must be to show that, however much Gladstone may yield, the Fenians are irreconcileable—"intransigentes."

We have been very anxious about our dear Lady

Muriel, who must be in a sad state of agitation and 1882.
anxiety about her husband; and, knowing that her
constitution is very delicate, we have much feared
the effect such a shock must have on her. We have
however heard from Cissy, and from Mrs. Wilson,
who have seen her in London, that she is well in
health, though of course very anxious and very much
distressed.

————

May 15th.

Yesterday I had the pleasure of receiving a letter
from Lady Muriel from the Vice Regal Lodge at
Dublin; a very sweet letter, as coming from *her* it
was sure to be, and as comfortable as could be at
such a time. She crossed over with Lady Spencer
and Mr. Trevelyan, on the 10th. She says they had
a perfectly good passage; they found Lord Spencer
and Mr. Boyle looking very pale and much fagged
(as well they may!)—but that all of them are very
well guarded and protected.

We returned the day before yesterday from Mildenhall, where we had spent between two and three
days, having gone thither on Wednesday. We were
most fortunate in the weather, which was perfectly
beautiful during the whole time, and we were able
to take a satisfactory survey of the new plantations,
which are doing very well, and some of which are
now sufficiently grown to make a comely appearance. The house and garden are in the most
perfect order, under the excellent care respec-
tively of Agnes Fincham and of Elmer.

We had much pleasant and satisfactory conver-

1882. sation with the admirable Vicar, Mr. Livingstone and his wife, as also with Mr. and Mrs. Robeson, whom we unexpectedly found their guests, and also with Mr. Lott. Mr. Livingstone and Mrs. Robeson are particularly agreeable.

_____

May 21st.

No trace yet of the murderers of Lord Frederick Cavendish and Mr. Burke — a strong proof how deep and widespread is the conspiracy of murder by which that deed was effected. A loud pretence of horror, yet, it would seem, a unanimous agreement to screen the murderers. I hardly believe that Albania or any part of Turkey, or of Europe, can be in a worse state, politically and morally, than Ireland. It seems hopeless. The Government has at last been roused to something like energy: the measure which they have brought in seems stringent and severe enough, but though it has passed its second reading by a large majority, I fear the Home Rulers will be able by their opposition in Committee, to delay it so much as to diminish its efficacy. I hope the Government will continue firm in carrying out its strongest provisions. It will indeed require much courage, in the Judges, and in all who have to carry it into effect.

_____

May 22nd.

News that the Duke of Grafton has died in

London, of typhoid fever. I am very sorry. He 1882.
was a very amiable and good man, and always very
friendly to *us*, though the poor Duchess's unfortunate
state of health has prevented them from doing anything for the society of the county.

The weather this month, hitherto almost continually fine and dry and bright, indeed brilliant ; of the
last eleven days, ten have been entirely without
rain.

—— ——

May 26th.

We came up yesterday to London, Minnie Napier
with us, and are settled at 48, Eaton Place for " the
season." As for myself I must repeat what I said
last year, of my reluctance to leave Barton at this
beautiful time of the year, and when the gardens
and pastures and woods are in their glory. Owing
however, to the remarkable mildness of the winter,
and indeed of the season hitherto altogether, the
periodical phenomena, such as leafing and flowering,
have all been very early this year, so that the
characteristic beauties of the *spring* are almost over
and those of *summer* rapidly becoming established.
The blossoms of the lilac and of the common horse-
chestnut are withering, those of the apple, the
cherry, the bird-cherry, the tulips, the narcissi, quite
past ; while the thorns white and red, the red horse-
chestnut, and the laburnum are in their glory. The
exquisite variety of delicate tints in the young
foliage is passing away, and the deeper and more
uniform greens of the summer are rapidly gaining
ground.

1882.　Very heavy rain yesterday morning, beginning before we left home, and continuing till towards 1, p.m.

_____

Yesterday was the 38th anniversary of our marriage, a blessed day which I can never remember without the liveliest emotions of gratitude to the Almighty for the innumerable blessings bestowed on me, and especially for the gift of a most precious wife.

Fanny has had a charming letter from Lady Muriel, who describes very pleasantly the life she leads with Lady Spencer and Lady Sarah, who seem to be making it as agreeable for her as it can be in such times. All is now apparently and out-wardly quiet, but she sees very little of her husband who is busy every day, and all day long. She describes the views about the Vice-regal Lodge and the Park, as most lovely—the lights and shades on the mountains quite exquisite.

_____

Spent an hour yesterday, much to my satisfaction, in the gallery of minerals in the Natural History Museum, Cromwell road.

This gallery has been arranged and opened to the public since last year. The collection of Minerals is magnificent, both in the beauty of the specimens and in the number of localities from which they are derived, illustrating to a great extent the geographical distribution and modes of occurrence. I noticed very numerous and fine specimens of native silver and native copper from the country about Lake Superior—many of them in very large and remarkably distinct crystals ; native gold from a vast number of localities, and among these some specimens from Gongo Soco in Brazil almost exactly similar to those I have, only larger. Also most magnificent crystals of sulphur, some from Conil in Spain, and others equally fine from Girgenti. The minerals in this collection are arranged according to a system which appears to me rather complicated and difficult, and would require a good deal of study. As far as I understand, it is *primarily* a chemical arrangement with a subordinate classification, according to the *crystallographic* characters—rather puzzling. The labelling however, is very clear and careful, and the hand catalogue and index to the minerals are useful.

June 2nd.

We went to see the exhibition of the Institute of Water Colour Painters : very much pleased with it.

June 3rd.

We yesterday visited Bull's nursery garden to see his exhibition of orchids—wonderfully beautiful. I have long been used to see this family of plants displayed in great beauty and variety at the flower shows of the Horticultural and Botanical Societies ;

1882. but this exhibition at Bull's far excels everything of the kind I have ever seen, in the multitude and mass of the lovely objects brought together, in the number of genera and species as well as of individuals, in the variety of forms, and in the beauty of colours.

After Bull's orchids, we visited *Rosa Bonheur's* picture of "The Lion at Home," a really noble painting. It is, to my thinking, the finest work by Rosa Bonheur, and decidedly the best representation of lions that I have ever seen. The lion is represented in the characteristic attitude of stately repose—"*a guisa di leon quando si posa,*" his head raised as if watchful; the lioness beside him turning her head round; three little cubs beside her. The lion especially, is magnificent.

Garibaldi died yesterday in his island of Caprera, having, I think, outlived his powers and influence, if not his fame.—Yet he was a man who ought by no means to be forgotten. He was not a completely and consistently great character, like Washington and the 1st William of Orange; but he had very great and splendid qualities, and performed very great—indeed one may say wonderful—actions. The Italian nation, above all, owe him an endless debt of gratitude, for it was mainly through his genius, patriotism and heroism, that it became established as a nation with its present extent of territory.

———

June 5th.

Mrs. Henry Talbot (Julia Berners) came to

luncheon with us. She always was a great favourite 1882. of mine, and I now think her more charming than ever. I am very sorry to have so rarely a chance of seeing her, as her husband has the charge of the military prison at Dublin.

————

June 6th.

A second visit to the Mineral Gallery. Splendid specimens of Chalcotrichite (capillary Red Copper) from "Fowey Consols" mine, St. Blazey, Cornwall. Corundum, a very large series of specimens from many localities, illustrating all the gradations, from the coarsest massive opaque varieties to the perfect Ruby and Sapphire.

————

June 7th.

I was mistaken, I see, in supposing that Garibaldi had outlived his influence ; at least, the news of his death seems to have created a great deal of excitement in Italy, and not in Italy alone.

————

June 12th.

A visit yesterday from Leopold and Mary Powys, very lately returned from Egypt. They went as far as the 1st Cataract. Leopold seems to have enjoyed the expedition amazingly, and the climate did his health a great deal of good. He describes the abundance and variety of birds, all along the Nile valley, as something astonishing ; he has made a fine collection, including several rare species : and

1882. he says that if he had shot for the sake of slaughter
(as many Englishmen do) and not for scientific
collecting, he could easily have killed thousands.
The plagues of Egypt, he says, are:—frogs (the
noise of which is incessant and most wearisome),
flies (not mosquitoes nor biting or stinging flies, but
teasing in the same way as they do in the South of
Europe, but much worse), and sandstorms.

He says that the Khedive is the best man in
Egypt,—that the country never was so prosperous
or so well governed as it is now, and that the great
reason for these machinations against the Khedive
is that he has protected the peasantry against the
oppressions of the officers and underlings civil and
military.

L. Powys believes that the disturbances in
Egypt have been secretly instigated by the Porte.

He says that it is very painful to an Englishman
to hear the way in which England is now every
where spoken of.

Powys says he found it absolutely necessary, in
Egypt, to shave his face smooth, because his
whiskers and beard became absolutely loaded with
sand.

June 13th.

Mr. Maskelyne (whom I met yesterday at the
Athenæum, where there was a ballot), says that the
systematic arrangement of minerals in the new
Natural History Museum is founded on that of
Gustav Rose, which is followed in the Museum of
Berlin. The double system of arrangement—by

the chemical composition and the crystallography— 1882.
is (he admits) difficult, but he thinks it the most
natural. He believes this (the British Museum)
collection of minerals to be the finest in existence :
that of Berlin the second ; that of Vienna was the
finest, but has not been so well kept up. The
Turin Museum, he says, is matchless for the
minerals of the Alps.

June 19th.

We returned to-day from a most agreeable visit
to our dear friends Kate and Charles Hoare, at
Minley Manor, about 4 miles from Sandhurst: a
place which they have hired from its owner Mr.
Raikes Currie. We went thither on the 15th, and
spent our time delightfully : the place and neighbourhood are beautiful, the house excellent; our
dear friends as kind and cordial as possible, and
agreeable in every way. Dear Kate as charming as
ever; she and her husband seem as happy in each
other and in their fine children, as ever we could
wish them to be.

The guests we met at Minley Manor were:—Mr.
John Marsham (brother of Lord Romney) and his
wife and daughter, Lady Harriet Fletcher, Miss
Caroline Hoare and my friend Mrs. Martineau—all
very pleasant people.

William Napier and two of his daughters, Colonel
Curzon, Charlie Napier and Arthur came to dinner
there on the 17th.

Minley Manor is a large house, quite modern ;

1882. the situation beautiful, on the southern brow of the extensive table-land of Bagshot sands called the Hartford Bridge Flats, commanding an extensive view to the S. and S. E. over a finely varied country to the high grounds beyond Aldershot and towards Farnham. The slopes descending from the edge of the table-land are beautifully formed, partly grassy and partly clothed with woods of pine and fir and birch, and through these woods (which were mostly planted by Mr. Raikes Currie) there are delightful walks and drives in many directions. The soil, being sand mixed with decayed vegetable matter, is the true "heath soil," naturally clothed with Ericæ, and perfectly suited to the growth of all that family of plants, and also of Coniferæ. It is very agreeable to see Rhododendrons and other beautiful shrubs flourishing in full vigour, as if wild, in the midst of the woods, among Scotch Pines, Heaths and Bracken. The Rhododendrons, indeed, are somewhat beginning to pass off, but the Kalmias (K. latifolia, which I think one of the most beautiful of plants) are in glorious vigour and profusion of blossom. A variety of exotic Conifers are scattered here and there in the woods, many of them very fine specimens; two Wellingtonias (Sequoia gigantea) near the house, are very large and fine, and another, standing in the wood, on a point commanding an extensive view, is decidedly the most superb specimen of the kind I have ever seen. Old trees there are few—only here and there a fine Beech.

On the way from London I observed in some

places where there were oak trees, that their leaves 1882. were shrivelled and blighted as if scorched by fire. I had heard of this, and it was said to have been the effect of the storm on the 29th of April. Our dinner party:—Lady Florence Barnardistone, the Yeatmans, the Leopold Powyses, the Lynedoch Gardiners, Mr. Gambier Parry, &c., &c.

June 22nd.

We had lately the news of the great fire at Bury, which caused a deplorable deal of loss of property ; happily, no loss of life, though several persons had a narrow escape. The greatest part of one side of the Abbeygate Street seems to have been destroyed, and that street was for a time rendered impassable by the ruins. There seems to be no doubt that it was purposely caused by a scoundrel with the object of cheating the insurance office.

June 23rd.

We went (Lady Grey with us), to see *Rosa Bonheur's* Lions (see June 2nd), and *Munkackzy's* picture "Christ before Pilate"—the latter a powerful but unpleasing picture.

Our dinner party :—the Albert Seymours, Lord Talbot,* Sir Robert and Lady Cunliffe, Lady Grey, Minnie, John Herbert, and Leonora.

June 24th.

I hardly remember a time when the aspect of

* Lord Talbot of Malahide.

1882. public affairs was more gloomy and alarming than
at present. Anarchy in Ireland—anarchy in Egypt
anarchy in the House of Commons ; so the state of
affairs has been described, and I think truly.

Nowhere any confidence in anybody or in any
thing—nowhere any feeling of security. The state
of Ireland is horrible, and it is believed that the
secret organization of murder in Ireland is in connection with the secret organizations on the Continent—the Nihilists and others. In such a state of
things, what can we say but—" Lord, have mercy
upon us."

—————

June 26th.

Our dinner party :—Lady Louisa Legge, the
Rayleighs, the Leckys, Augustus and Rachel Vernon
Harcourt, Sir F. and Miss Doyle, Sir Lambton and
Lady Loraine, MacMurdo, William Napier.

—————

June 27th.

We dined with Lord Chief Justice Coleridge; met
the Archbishop of Dublin, the Bishop of Truro, Lord
and Lady Sherbrooke, Sir Henry and Lady Holland,
Mr. and Mrs. Montgomerie, and several others ;
altogether an interesting party—Miss Coleridge pre-
siding. The house is beautiful with works of art.

—————

June 28th.

We dined with Katharine ; met the Hookers, the
Maskelynes, William Nicholson and his daughter ;

Leonora and her two daughters — an agreeable 1882.
party ; afterwards, a crowded evening party.

———

Jnly 3rd.

We are just returnèd from a visit to Lady Grey at
Fairmile near Cobham ; an exceedingly pleasant
visit ; we went thither on Thursday the 29th of
June, and have been fortunate in fine and warm
weather (for summer has at last begun), which
has allowed us to enjoy the delightful country.
Lady Grey is most cordial and friendly, and she has
such a fine mind, such nobleness of character, and
such high cultivation, that her company is both
interesting and improving. She is one to be both
admired and venerated.

Lady Grey's house is not a large, nor yet a very
small one, but of a comfortable size ; not splendid
but most eminently well arranged and comfortable
— indeed to me it appears almost absolutely perfect
(for its size) in all its arrangements. It was not
originally built by her, but she has altered it so
much since buying it, that I have no doubt, all
the merit of its construction and adaptation belongs
to her. It has a small but pretty garden, opening
on the old Portsmouth high road, between Esher
and Cobham, and very near Claremont ; the sur-
rounding country very pretty and pleasant, especially
from the numerous open spaces, heaths, and commons, which are so exhilarating in comparison with
the monotonous hedge-and-ditch country so prevalent in England. I had two delightful walks with
Lady Grey, over the nearest heath and through two

1882. woods, the one chiefly of fine red-stemmed Scotch
Firs, and enclosing a large pond or mere with
swampy margins, which afforded a variety of marsh
plants. In particular, here grew abundance of
Hypericum elodes, a plant which I had not seen
growing since (I think) 1832 ; it is not yet in flower,
but formed whole beds on the margins of the pond,
and quite in the water. The other wood lies
between the common and the river Mole, sloping
down to the bank of that sluggish little river ; in the
higher parts it is composed chiefly of Scotch Firs,
with an undergrowth of bracken ; lower down the
soil is damp, and the growth more varied ; here
I observed many interesting plants, especially Scir-
pus sylvaticus (which I had not seen alive for
many years), Œnanthe crocata, Lysimachia ne-
morum, and Corydalis claviculata. The Foxgloves
in this wood were quite extraordinary, both for
abundance and height ; some of them must have
been at least 7ft. high.

In this neighbourhood, the oaks and the horse-
chestnuts are putting forth abundance of fresh,
vigorous and healthy coloured shoots, the previous
growth of the year having been almost universally
blighted and seared by the storms. We met at
Lady Grey's some very agreeable guests :—On the
29th, Matthew Arnold and his daughter ; on the
30th, Mrs. Charles Buxton, John Carrick Moore
and his daughter. Matthew Arnold in great force,
full of good talk, very agreeable ; his daughter a
nice lively, pleasant girl ; Moore (who is some years
my senior) very agreeable ; Mrs. Buxton also.

July the 1st, Lady Grey went with us to call, first 1882.
on John Moore, at Brook Farm, where we saw
original portraits of Sir John Moore, Sir Graham
Moore, Sir Ralph Abercrombie, and Lord Erskine,
and also an oak tree planted by Sir John Moore just
before he set out for his fatal campaign in Spain.
Secondly, we visited Mrs. Charles Buxton, at Fox
Warren (such I understood is the name), a handsome modern house in a very fine situation. I find it
difficult to make out the topography of that
neighbourhood, and to trace on the old ordnance
map.

Mrs. Buxton's house stands on the brow of a high
and very steep hill, commanding a very extensive
view of a beautifully varied country, with much
wood and much heath ; noble old Scotch firs on the
brow of the hill, and a very remarkable tree of the
same kind on the summit. Mrs. Buxton's cockatoos
at liberty and flying about.

———

July 4th.

Our dinner party :—Lady Alfred Hervey, Lady
Rayleigh, Sir Francis and Lady Boileau, Mr. and
Mrs. and Miss Bonham Carter, the Leonard Lyells,
Kinglake, Mr. and Mrs. Roundell, Mrs. Laurie,
Hugh Hoare, Edward, Sir Edward Ducane, and
Cecil.

———

July 6th.

Our dinner party :— Captain and Lady Edith
Adeane, M. and Mdme. Ernest de Bunsen, Mrs.

1882. and Miss Galton, the MacMurdos (three) Mr. and
Mrs. Martineau, Reginald Talbot. Miss Sulivan,
Edward, &c., &c.

———

July 7th.

We went to see the chapel at the Wellington
barracks—very elaborate and very beautiful, in the
most highly ornate style, but the richness of the
coloured windows too much diminishes the light and
makes it difficult to see the details.   What are
most interesting are the inscriptions everywhere in
commemoration of the brave men—officers and
men—of the Royal Foot Guards, who have served
their country faithfully and died in her service.

Afterwards I went into the Royal Academy
Exhibition, and saw it tolerably well; thought it
rather mediocre; many good pictures, but few if
any striking ones that left a strong and lively
impression on the mind.   Some capital portraits.

———

July 9th.

Louis Mallet thinks very despondingly (as I do)
of the state of the political world.   He evidently
thinks that Gladstone's government has been a
complete failure, especially as to Ireland ; and if, as
is most probable, the Conservatives come into
power, they can hold it but for a few years, and
then must come the dominion of the thorough
Radicals.   He thinks that it will not be possible
long to prevent the Irish from gaining their " Home

Rule:" they have found out the way of making 1882.
themselves utterly intolerable in the House of
Commons, and stopping the progress of all business
except the Irish, so that it will be found inevitable,
before many years are past, to let them have a
separate Irish legislature, else all the business of
England and Scotland will be brought to a complete
standstill.

Mr. Lecky said to me—I wonder how much
longer England will tolerate the complete stoppage
of all its business in the House of Commons.

Louis Mallet says (what surprised me), that the
voyage to India round the Cape is now made by
steamers in a time longer by only 4 days than that
in which the voyage by the Suez canal is performed.

July 11th.

We visited old Lady Lilford, and had a long and
very pleasant conversation; she is remarkably well
informed and agreeable, and always a warm friend
to us.

July 12th.

We went to the British Museum (the one in great
Russell Street, not the Natural History one), and
saw Mr. Bond the head librarian, and Mr. Reid the
keeper of the Prints, to whom we had introductions
from Lord Walsingham. Both were very kind and
courteous, and showed us some interesting things.

The news from Alexandria terrible : — the desertion of the city by the army, its abandonment to a horde of released convicts, the conflagrations and the atrocities to which it appears to have been given up.

Not having followed the newspaper details of this Egyptian business in its earlier details, I do not feel myself qualified to judge whether our recent violent proceedings there are justified by necessity; but at any rate I am sure that it is a painful and deplorable necessity. The amount of the slaughter caused by our bombardment does not appear to be known, but I fear it must have been very great.

The Prevention of Crime (Ireland) Bill, has at last become law — certainly not too soon—two months after its introduction. It is no doubt a strong remedy; I wish it may be efficacious. We dined yesterday with the Leckys; met Sir Henry and Lady Loch, Lady Strangford, Lord and Miss Talbot, Mr. Gibson (the leading Irish Conservative in the House of Commons, a very able man), Mr. Hamilton Aidé and Mr. James, the American novelist, Lady Loch (whom I took into dinner) is very pretty and very agreeable.

Mrs. Lecky said to me that she was very glad and rather surprised, that the Americans are not angry with Mr. Lecky's History of the War of Independence, in which he has shown the behaviour of the Americans in a less favourable light than that in which it has usually appeared. They have seen,

she says, a great many American newspaper reviews
on the book, and all are favourable.

——— ———

Our dinner party : — Lady Winchelsea, Lady
Evelyn Finch-Hatton, Mr. and Mrs. Coore, Lord
and Miss Talbot, Matthew Arnold, Mr. and Mrs.
Longley, Katharine Lyell, Mr. Sinclair, Clement,
&c.

——— ———

Bright has resigned ; quite right, and certainly
to his honour. It would have looked very ill for
him if he had continued a prominent member of a
Ministry which has now engaged actively in a war.

——— ———

48, Eaton Place.

We were present yesterday at the marriage of
dear Emily (Mimi) MacMurdo to William Bruce,
in Fulham Church. A beautiful couple they are,
and seem peculiarly suited to each other ; God
grant them long lives of married happiness, for with
His help I am sure their lives will be a blessing to
all connected with them. The scene in the Church
was very pretty. There were about 42 assembled
there—of the bridal party, I mean—and nearly as
many at the "breakfast" at the MacMurdo's : all,
or almost all, related or connected with the two
families. Augustus and Rachael Vernon Harcourt

1882. (whom I was extremely glad to meet), had come all
the way from the Isle of Wight, and Harry Bruce
all the way from Wales, to attend the wedding.
Norah Aberdare, whom I was agreeably surprised
to see, because she has been very much out of
health, was present in the church, but not at the
breakfast; she does not, however, *look* ill.

We went home to Barton on the 20th of July,
and remained quiet and happy there, enjoying the
tranquil beauty of the country, till the 3rd of this
month, when we came back to Eaton Place, solely
in order to be present at dear Mimi's wedding. We
were very quiet at Barton, but not quite alone, for
Arthur joined us on the 24th, and Katharine on the
29th, but we did not see much of our neighbours.
On August the 1st, Mrs. Horton sent Fanny the
good news of Jane Broke's engagement to be
married to Mr. Saumarez, the eldest son of Lord
De Saumarez. He is very well spoken of, and
everybody concerned seems very much pleased, and
I am very glad, for she is an excellent as well as
very clever young woman. Since we came to town
we have bought wedding presents for her and for
Emily Egerton, a charming girl and a great
favourite of mine, who is engaged to Mr. Fielding,
a clergyman and a brother of Lord Denbigh.

———

August 7th.

During the fortnight which we spent at Barton,
the weather was variable; never very bad, but
scarcely very fine for a whole day together, and

never very hot. Such too it seemed to have been 1882.
on the whole through most part of July, and therefore the hay harvest was very late and very slow,
being interrupted, more or less, almost every day.
Scott, however, succeeded, by constant attention
and watchfulness, in making up a fine stack in
excellent condition at our stables : but our hay-
making was not finished till July 26th.

In coming up to London on the 3rd, I observed
haymaking still going on in several spots, while
at the very same time in other fields, wheat was
cutting. We had the good news that Arthur had
passed his concluding examination at Sandhurst ; so
that he is now in effect a commissioned officer in
H. M. service, and only waits to be appointed to
a vacancy in some regiment. He is extremely eager
to be employed.

August 9th.

We returned home to Barton yesterday; Leonora,
her daughter Dora, and Arthur with us. The
weather splendid, and the wheat harvest going on
gloriously. The wheat and barley fields make really
a beautiful show.

August 15th.

Weather delightful, almost ever since we returned
home : the 10th indeed was a cloudy and dull day,
though not wet, but with this exception the weather
has been (until to-day) everything one could desire,
and the harvest has gone on as well as possible.

1882. Much wheat, I hope, has been carried in the last 24 hours, and most, if not all the wheat in the parish, has been cut since the beginning of the month. To-day however the weather has been unsettled, there was rain in the morning, and a very heavy shower in the afternoon.

August 16th.

Signed the deed of Presentation of the living of Barton to Mr. Harry Jones;—Mr. Percy Smith having decided to resign it, on account of his wife's health, and having obtained a permanent chaplainship at Cannes.

August 25th.

Weather lamentably changed since the 20th, has become cold, rough, blustering and unsettled, with storms of rain now and then. The rainfall in the 24 hours, ending 9 a.m., on the 23rd, was 0.47 inch. Happily, in this parish, and those immediately around us, I believe, the wheat has almost all been carried, in good condition. The barley is, for the most part, still out, but even as to this, Scott appears to be somewhat hopeful.

August 29th.

Mr. and Mrs. Percy Smith had luncheon with us, and took their leave. He preached his farewell sermon (on his departure from Barton) yesterday afternoon, and I hear that the Church was crowded, and every one was in tears. Unfortunately we were not there, having been at Church in the morning. 1882. It is not surprising that the Barton people should be touched with sorrow at parting with Mr. Percy Smith, for I have seldom known a more amiable man, or one more full of human kindness, mildness, and good will towards all whom he had to do with.

September 3rd.

Mr. Harry Jones (though not yet regularly *inducted* into the Vicarage of Barton) read the service and preached in the Church this morning; delivering also a very good prefatory address to his new congregation.

September 4th.

Very agreeable friends staying with us from time to time since our return from London; Leonora and her two very nice and clever girls ; the MacMurdos (I always find them very pleasant, and his talk full of instructive as well as amusing matter) ; Mrs. Douglas Galton (very agreeable) ; Lady Louisa Legge (very lively, clever and amusing, and very good natured) ; Clements Markham (full of various knowledge and anecdote, agreeable as usual); Archdeacon Chapman (enlightened, sensible and good, a most estimable man).

September 9th.

Arthur has had an official intimation of his appointment to a commission in the Highland

1882. Light Infantry Regiment. (I am not sure that I am perfectly correct in its new-fangled title) ; its two battalions were formerly the 71st and 78th Regiments, and he does not yet know which he is to join. He is, of course, highly delighted, and eager to be sent out at once to Egypt ; but it is perhaps more likely that he will have to join the depôt—at first at any rate. If God is pleased to spare him, I have no doubt he will make an excellent soldier. But I cannot help having considerable uncertainty and apprehension as to the course the war may take —and fear that it may prove much more difficult and sanguinary than those who direct it seem as yet to expect.

————

September 10th.

Most beautiful weather these last four days ; brilliant sunshine, with a fresh, not strong, rather sharp, exhilarating, easterly breeze. Delightful harvest weather. The wheat harvest is now, I think, quite finished in this parish, and in most of the neighbourhood, and has been got in in fine condition. The barley harvest is not everywhere finished, but our two principal farmers, Mr. Phillips and Mr. Cooper have got all theirs in and I should think that with this fine weather, all the crops are likely very soon to be stacked.

Wasps have become suddenly very numerous in the last few days ; before I had hardly seen one of them.

————

September 14th.

Yesterday evening we had the news of the

battle and the great victory of our army at Tel-el- 1882.
Kebir in Egypt.   It is indeed a splendid action and
a splendid victory, and does great honour both to
the general and the army, and I hope it will lead to
a speedy peace.

- - - - -

September 15th.

The hope with which I ended the last entry in
this journal has been fulfilled sooner than I
expected.   The great victory at Tel-el-Kebir has
been followed in an astonishingly short time by the
surrender of Cairo, the capture of Arabi and the
other principal chiefs, and in fact, as it appears, the
total collapse of the insurrection.   I gladly retract
all the expressions of uncertainty and apprehension
which I wrote on the ninth ; every feeling of that
kind has been dissipated by the skill and ability of
the general, and the splendid fighting of our army.
The war seems to be really at an end, and all that
remains to be done falls within the domain of
diplomacy.   I wish that politicians may not mar the
effects of all that has been done by arms.

- - - - -

September 17th.

At morning service in Church, Mr. Harry Jones
gave us a singular and striking sermon—or rather
lecture—on the events in Egypt, having himself
visited that country a few years ago. His view is
that this war is a sort of crisis in the great struggle,
which has been gathering and approaching for some
years, between Islamism and Christianity.

1882.   It certainly does seem, that for some time past, several Mahomedan nations have been showing a more haughty and aggressive spirit than before.

* * *

September 18th.

William Napier writes of the war in Egypt, that he can call it " nothing less than wonderful ; it was " well planned, well executed, and well followed up. " Sir Garnet has, indeed, deserved well of his " country." Truly, I think no one can henceforth sneer at his generalship.

LETTER.

Barton Hall, Bury St. Edmund's,
September 19th, 1882.

My Dear Katharine,

Arthur has just left us, to spend a few days at Rose Bank and in London to order his uniform and equipment preparatory to joining his regiment, for he is now actually a commissioned officer in H.M. service. He is gazetted to the 1st Battalion of the Highland Light Infantry—the old 71st, and is to join at Glasgow on the 29th of this month ; so there is not much time to be lost. He is in very good spirits, though (I think) a little disappointed that the war is over. Fanny, as you may imagine, does not share in the feeling of disappointment ; she is very much relieved in mind, having been excessively anxious while we thought

he would have to join the regiment in Egypt. It is 1882.
a really marvellous change since the beginning of
this campaign.

I do not know your opinion about the war, but I
hope you rejoice as we do in the brilliant success of
our country's arms, the admirable way in which
the whole thing has been managed, and the
rapidity of the result.

William Napier writes that he cannot call it less
than wonderful, well planned, well executed and
well followed up. I think nobody in future can
sneer at Sir Garnet. But now comes the question
—what next? I hope we shall not throw away by
diplomacy all that has been gained by fighting.

Our new Vicar, Mr. Harry Jones, gave us last
Sunday, a striking and interesting *lecture*, rather than
a sermon, on Egypt and the war, having himself
been in that country a few years ago. He represents
this war as a crisis in the great struggle which
he believes has been gathering and drawing nearer
for several years past between Islamism and
Christianity. If it be so, I rejoice that England
has stood forward in the front for Christianity ; but
it does not look as if the wretched Egyptian serfs
who have been forced into the army were capable
even of fanaticism.

Fanny's journal, I know, tells you all our news,
and indeed I do not know any. Dear Cissy and
Emmie arrived yesterday (19th) and for a few days,
I believe, we shall be alone with them. I am reading Mrs. Kemble, and am not now far from the end
of the book ; I have found a great deal in it both to

1882. entertain and interest me, but I confess I am growing rather tired of the third volume, perhaps because there is more of melancholy in it than in the others. Altogether the book, though it gives a high idea of her intellect, does not impress me so much with a sense of her amiability ; however I may be unjust to her, from not knowing enough of her personal history, and of the provocations she may have received. There is rather an uncomfortable mystery about all this, and I can only see vaguely that she seems to have been very unfortunate. Our friend Lady Grey is very fond of her, which is a very strong point in her favour. Her anecdotes and sketches of London people are uncommonly entertaining, and I should think very true.

Our garden has still a great deal of beauty and in the hothouse the Allamanda and Bougainvillia are in full blossom. A few red leaves are appearing on the Liquidambar, and yellow ones on the Horse-chestnuts.

Give my love to Rosamond, and (if this letter finds you at Shielhill) my kindest remembrances to the Miss Lyells.

With Fanny's best love,
Believe me ever,
Your loving brother,
CHARLES J. F. BUNBURY.

JOURNAL.

September 20th.

Arthur went yesterday to London to prepare his

outfit—uniform, etc., for joining his regiment. He 1882.
is gazetted to the 1st Battalion of the " Highland
Light Infantry," formerly the 71st, and is to join at
Glasgow. I understand it is a very fine regiment;
it was a famous one in the Great War.

***

September 23rd.

Arthur returned from London, having leave of
absence till November 9th.

***

October 2nd.

The house full of company from the 25th, to the
30th of last month:—William Napier, Lady Grey,
Lord Talbot de Malahide, Miss Talbot, the Dean
of Ely, Mrs. and Miss Merivale, Annie Campbell,
and her sisters Finetta and Griselda, Guy Campbell,
young William Parker, and at last, Edward. W.
Napier looking very well, wonderfully young and
active for his years, and always delightful: moreover
one of the very best men I know. Annie Campbell
lovely and charming.

***

October 3rd.

Dear Cissy and Emmie with us from the 19th to
this morning. I have omitted also two Wilson
girls—Agnes, delightful; and Ida, uncommonly
pretty. All the Wilson girls are pleasing, very
pleasant to look at.

***

October 7th.

I had my Barton rent audit yesterday—it was satisfactory and comfortable. The rents, of course, considerably lower than they were formerly, as I had found it necessary to agree to a reduction in the *bad years*, the effect of which has by no means passed away, but there were no defaulters, and I have the comfort of having no farms left on my hands, and of finding the tenants not merely civil, but expressing themselves as grateful for my sympathy.

The harvest of this year it is admitted is good (that of wheat especially) but the markets are absolutely glutted, so that it is difficult to dispose of the produce.

My Mildenhall rent audit took place on the 2nd and (as Scott informed me) passed off very well on the whole. " The tenants," he said, " all spoke of " good crops and bad prices, but were much more " cheerful and hopeful than at this time last year."

October 13th.

Another large party in the house :—Leopold and Lady Mary Powys, Mr. and Mrs. Sancroft Holmes, Frederick and Fanny Jeune, General and Mrs. Ives, Ellen and Ethel Wilson, and now, lastly, Lady Susan Milbank with her two children. Dear Minnie has been with us all along, and I hope will remain till next month.

The Holmeses are a remarkably agreeable and intelligent young couple, whom I think, I have mentioned before. Since they were here last year they have visited Ceylon, where (I think) they have property ; and we had much talk about tropical nature, which they seemed to have greatly enjoyed ; *she* especially, as she has a great love of plants, and is a good observer, though without accurate botanical knowledge. Of Leopold and Mary Powys, I have often spoken before. Leopold is always agreeable, and I especially enjoy a talk with him on natural history, to which he is devoted. We are very fond of them both.

"Fritz" Jeune, a very clever and agreeable young man.

———

October 14th.

Poor Arthur has been completely laid up, all this week, in consequence of a most unlucky accident, or complication of accidents, and I fear it will be some time before he can be quite recovered. It was I think, on the 7th, that while shooting at Freckenham, he gave himself a sprain in the groin, in leaping, which proved more serious than he was aware of, and which has ever since disabled him from walking. The very same day he caught a violent cold, which took a feverish turn, so that for several days he was confined to bed in a state of great discomfort. Now, happily, he has quite shaken off the cold and fever, and is able to lie on his sofa, but the injury caused by the sprain is not cured, and he is not yet able to walk with safety.

———

October 21st.

Clement came to us yesterday— very lately

1882. returned from America. He had taken a long journey into the interior, past the great lakes, to the N.W. into the district now called Manitoba, in the former Hudson's Bay territory. He was delighted with the lakes, but found the scenery of the prairies monotonous and dreary in the extreme; nothing but a uniform surface of stunted grass, without a tree or a rock; no variety, but an abundance of sunflowers, and others resembling them. No buffaloes now to be seen—all driven away by railroads and advancing population.

Niagara fully up to his expectation—impossible to be disappointed with it. What impressed him most was the mass and especially the depth of the falling water, preserving its fine, clear, green colour, almost to the bottom.

Scenery of the Hudson river very beautiful.

Prevalence of wood in the building of country houses.

He was very much struck with the magnificence of the public buildings (the Capitol at Washington for example) and the great scale of the public institutions.

October 25th.

There was yesterday a really tremendous storm of wind, which raged through nearly the whole day, to evening with furious rain, and at one time in the afternoon, real *snow*. It has covered the ground with broken branches, and twigs as well as leaves; but as yet, I find only two whole trees destroyed; a

large, tall elm (standing at the corner where the 1882.
road to the north lodge enters the north grove) which
is snapped off close to the ground, and in its fall
has smashed a good sized oak. The trunk of the
elm was quite decayed in the centre, though
apparently sound outside.

A great elm near the flower-garden gate, and an
oak near the further corner of the pleasure ground,
have each lost a large branch. I have not yet
heard of any other damage. The storm seems to
have been very general and very destructive in many
parts.

I am thankful that no damage of any consequence
has been done in the arboretum.

————

October 26th.

Thermometer in garden down in the night to 28
deg. ; the first frost since the spring.

————

October 30th.

The rainfall in 48 hours from 9 a.m. on the 27th,
to 9 a.m. on 29th—1·09 inches.

————

November 4th.

I received an interesting letter from Edward, from
Corte in Corsica (finished at Leghorn). He is
delighted with what he has seen. Corte, he says
is " an extremely picturesque old town ; in point of
" situation one of the most striking I have seen even in

1882. " the Apennines, and situated at the foot of an
" extremely fine range of mountains.   Indeed, this
" central range of the Corsican mountains is in some
" respects, one of the finest I know.   Though com-
" posed entirely of granite which often produces
" tame and uninteresting forms, they are peculiarly
" rugged, and broken into ranges of jagged peaks,
" among the wildest and most fantastic that I have
" ever seen, except the Dolomites: and as they
" rise to between 8,000 and 9,000 feet, they are
" high enough to be grand as well as picturesque."
" I have found that by following the high road from
" Ajaccio to this place, one passes through some of
" the finest scenery in the island, and one has the
" opportunity also of seeing one of the forests for
" which Corsica is famous; but with this I was
" rather disappointed.   I believe there are others
" much finer, as well as more extensive, but after
" all, a forest of Pinus Laricio is very much like a
" forest of Scotch firs.   The scattered trees on the
" outskirts may be finer (though I did not see any-
" thing very remarkable in this way) but as soon as
" they grow thick they become a mere mass of poles
" like any other pines.   There are very fine chest-
" nut woods in many parts along the slopes of the
" mountains; but a very large part of the island is
" covered only with what they call *maquis*, answering
" to the Italian *macchia*—a mass of brushwood, of
" arbutus, lentisk, cistus, &c.—such as one sees so
" much of in Sicily and some parts of Italy, and
" still more in Greece, but there is nothing in
" Corsica like the richness and luxuriance, as well as
" variety, of that in Greece.

"At Ajaccio one sees aloes and prickly pears and 1882.
" a few palms, but it is evident that they are all of
" very recent introduction."

"The backward condition of everything in
" Corsica is a curious contrast to the mainland of
" the Riviera : and though Ajaccio is far ahead of
" everything else in the island, it is still much in
" the same condition that such places as Mentone
" and San Remo were in the old days."

LETTER.

Barton,
November 4th, 82.

My dear Edward,

I was delighted to get your letter, finished
at Leghorn, and have read it with great pleasure ;
we had not known when you left England, and a
letter which I wrote to you on the 17th of last
month (to tell you of Arthur's accident) was
returned to me from your lodgings. Arthur's
accident might have been a very serious one, but
happily it has not turned out so bad as it was at
one time thought, and he is now nearly well. It happened on the 6th or 7th of last month when he was
out shooting ; in leaping he strained himself severely
in the groin, and was in consequence confined to his
bed for several days (with a feverish cold into the
bargain, which he had caught at the same time),
and to a sofa for a much longer time : but as I said,
he is now very nearly well, and hopes to be able to
set off next Wednesday to join his regiment at

1882. Glasgow on the 9th. Poor fellow, with his active habits and devotion to out-door pursuits, so long an imprisonment has been a severe trial to him, and very fatiguing also to Fanny, who was constant in her attendance on him.

Your account of Corsica has interested us much, and I am very glad you saw it so well; what you say about the mountains is quite in accordance with the idea I had formed of them from Lear's book, which we have. They must be very noble mountains indeed. I can easily believe what you say about the pine-forests, for one knows how apt all the pines are, when crowded, to run up into mere poles. MacMurdo told me that in the Himalaya he had seen even Deodars drawn up in this way. I am very glad you have been so fortunate as to weather, which seems to have been bad over so large a part of Europe. Here we have certainly had, in the last month, some remarkable falls of rain, and I suppose on the whole, more than the average, but we have also had some very beautiful days, and I think few *entirely* wet. I think on the whole the prevalence of high winds has been more remarkable here than that of rain, again and again it has blown harder than was pleasant, and on the 24th there was a really tremendous storm, but happily it did no serious mischief here—only blew down a tall old elm which was decayed at the root, and which smashed up a pretty good oak in its fall. I hope you will have good weather in Sicily.

We have had much pleasant company in the house in the course of October, but a mere list of

names would not be very entertaining, and besides
I cannot write more just now.   I have no news to
tell you.

  With much love from Fanny,
   Believe me ever,
    Your affectionate brother,
     CHARLES J. F. BUNBURY.

JOURNAL.

November 8th.

Dear Arthur left us this morning by the early
train for London, where he is to see some of his
family, to continue his journey to-morrow, and join
his regiment (the 1st battalion of the Highland
Light Infantry — formerly the 71st), at Glasgow.
He has, I trust, quite recovered from the effects
of his accident, is looking well, and is in high
spirits, though regretting to have missed the chance
of active service in Egypt.   May God bless and
prosper him.   I feel no doubt he will be a good
soldier.

November 9th.

We saw the *comet* this morning, getting out of bed
at 5 o'clock on purpose; saw it very well, the sky
in that quarter being quite clear and free from
clouds.   It is indeed a splendid one,—a beautiful
and wonderful object : almost finer even than the
comet of '58.   It is the fourth I remember to have

1882. seen, but of that of '43, I saw nothing but the tail, which indeed was a wonderful one.

———————

November 10th.

We had the comfort of hearing (by telegraph) of Arthur's safe arrival at Glasgow.

———————

November 15th.

Very pleasant company last week and until yesterday :—dear Kate Hoare, Mr. and Mrs. and Miss · Walrond, Miss Loch, Harry Bruce, the Loraines (for part of the time), and John Hervey, the Livingstones and Mr. Barber (one of the curates at Mildenhall) for the last day. Dear Kate as charming as ever, and as cordial and loving to us ; Miss Loch* (whose father, George Loch, was an old friend of Fanny and her family), a very pretty woman, and very intelligent and agreeable, as well as good ; Mr. Walrond I have often praised, and never too much : I delight in his company— he has so much intellect and knowledge, and at the same time so gentle and quiet, so completely free from the dogmatic and domineering tone frequent in official men. He has also shown us a great deal of kindness with reference to our nephews Arthur and Clement. His daughter Theodora is a very handsome and very pleasing girl. Theodora Walrond is not as beautiful as her lovely and charming mother, a dear friend of ours, whose

* Emily Loch.

death* we shall always lament; but she is very graceful, and has fine features, and she repeatedly reminded us of her mother: she is not quite 18.

November 16th.

The two dear little Seymour children, Sarah and Albert's children, Charlie and Bill, have been with us ever since the last day of October. They are charming little boys.

November 17th.

Dear Minnie and her two darling little grand-children left us, we are very sorry to part with them.

November 18th.

We had the great comfort yesterday, of hearing by telegraph of Arthur's safe arrival at Dublin, and that he had had a good passage across the sea. He will probably be quartered at the Curragh of Kildare.

November 19th.

Mr. Harry Jones drank tea with us. He is a very agreeable man, as well as a very good one, and an excellent preacher.

November 25th.

We received the very very sad news of the death

* She died in 1872.

1882. of our dear friend Edward Campbell.   He had
indeed been in bad health for some considerable time,
and not long ago was dangerously ill, so as to be
thought to be in immediate danger; but that crisis
apparently passed over, and only two or three days
ago Dr. Andrew Clarke pronounced him to be
comparatively well.   A sudden relapse I suppose
carried him off.

I grieve very much for him, or rather for the loss
of him ; on *his own* account we ought rather to
rejoice, feeling a sure hope and trust that he is gone
to a world of eternal blessedness, to be reunited with
his beloved wife and with the good who have gone
before him.   He was a most excellently good
man, and an uncommonly pleasant one; of a
gentle, loving disposition, with a great deal of
quiet, kindly humour; deeply religious without
sourness or asperity.

————

December 4th.

I must begin this new part of my journal, with a
very sad event, though one which has been for some
time foreseen and expected.   The Archbishop of
Canterbury is dead.   He is a man to be deeply
lamented by those who knew him, and a very great
public loss ; a most excellent and valuable man.
He was an old friend of Fanny's family.   Mr.
Horner knew him well when he was a schoolboy,
had a great regard for him, and used to tell him (in
joke) that he would be a Bishop some day.   He
never forgot this early intimacy with the family, and
was always very friendly and pleasant whenever we

met him.   His manner indeed was very pleasing ;
the obituary notice in *The Times* uses a very
appropriate expression in speaking of his " bland
geniality."   That article in *The Times* is altogether
a very good one, and does him real justice.

The trial of Arabi, which seemed likely to be as
long as that of "the *Claimant*," has come to a very
sudden and surprising, yet (to my thinking) very
satisfactory conclusion.   He has been induced by
his European advisers to plead guilty ;   he has
been sentenced to death, and the sentence has been
commuted into one of banishment for life to some
country out of Egypt.   Probably he will go to some
British colony or dependency.

––––––––

December 6th.

Fanny has had a most deeply and painfully
affecting letter from poor Mrs. Willoughby Burrell,
to whom she had written to condole with her on the
death of her eldest son, a little boy of six years old.
He seems to have been an uncommonly interesting
and charming child, and the poor mother is almost
heartbroken : the expressions of her grief are most
pathetic.

There has been, within the last month, a sad
accumulation of losses and sorrows among our
friends :—this case of the Burrells (I think) first ;
then Edward Campbell's death ; then that of
Thomas Powys, Lord Lilford's eldest son ; then the
Archbishop's ; and to-day we see in the newspapers
that poor Sir Francis Doyle's eldest son, a very fine
young man and good soldier, has died from the

1882. effects of the Egyptian campaign. A melancholy list of those mown down in so short a time. It is indeed as Longfellow says in his beautiful poem of " Resignation."—

> " The air is full of farewells to the dying,
> And mourning for the dead."

————

December 7th.

Another death—but not of a friend or even an acquaintance, but of one well known by his works— Antony Trollope. Some of his novels are delightful, especially (to my taste) "Barchester Towers," and "Framley Parsonage."

There was yesterday a great astronomical pheno- menon—the Transit of Venus—which we ought to have seen, if it had been visible; but there was nothing to be seen but one uniform, dull, impene- trable grey veil of cloud, extending across the whole sky, and continuing unbroken through nearly the whole day.

————

December 10th.

Ceylon, it appears, is to be Arabi's place of banishment; not at all a bad one, I should think, for one who has been used, as he has, to a hot climate. I am glad he is not put to death; indeed to do *that* would have been a shame, after it had been made so evident that he had been instigated and encouraged in all he did by the Sultan or his Ministers, and by a great proportion of the most important people in Egypt. I suppose that in dealing with Orientals, we must always expect treachery, and in dealing with the Sublime Porte we seem to have met with an ample allowance thereof.

LETTER.

Barton, Bury St. Edmund's,
December 12th, '82.

My dear Edward,   1882.

I will begin a letter to you, to be finished
when we hear (as I hope we soon shall), of your
arrival in London.   I am very sorry that your
journey home must have been made in such very
severe weather.   I hope with all my heart that
you may not have caught cold, but I am afraid,
at any rate, you must have suffered a good deal
from the intensity of the cold.   Winter has set
in very early and with uncommon severity ; the first
fall of snow was last Thursday, the 7th ; here it was
accompanied by a high wind, but nothing amounting
to a storm ; but in several parts of the country
there seems to have been a formidable storm with a
great snowfall—roads and railways blocked, trains
buried in the snow, and some lives lost.   The next
night there was another fall, and since then there
has been continued hard frost.

Yesterday and the day before were very fine calm
days, with bright sunshine, but hard frost all day
where the sun did not reach—the snow quite crisp ;
yesterday there was a fog in the morning, and the
trees were beautifully hung with rime frost   I hope
the winter will not be long as well as early, or it will
be very difficult to keep the labouring people constantly employed and paid.   In my letter directed

1882. to you at Marseilles, I told you of the death of dear
Edward Campbell, which was a deep grief to us.
Since then there has been quite an accumulation of
deaths and sorrows of our friends, or our friends'
children; the poor Willoughby Burrells (whom I
think you saw at Cannes in the early part of your
tour), lost the eldest boy of six years old, said to be
a most charming child. Then Lord Lilford's eldest
son died, a great friend of Arthur's, and of just
the same age; then the good Archbishop, a great
public as well as private loss; and lastly, Captain
Doyle, poor Sir Francis's son, one of the many
victims of typhoid fever, caught in Egypt. This is
a sad list. The Archbishop is a grievous loss—
a loss, as Mr. Harry Jones says, that at present
is quite irreparable.

(*Dec. 14th.*)—No news yet of your return; we
cannot help being anxious lest you should have
fallen ill in your journey through France in this
cruel weather. But we must hope that we shall
soon hear from you.

I was very much interested by what you mention
in your last letter of your observations in Sicily,
I am very glad to hear there is so much improvement in that island, and especially to hear of the
improvement (though I shall never profit by it),
in roads and accommodations for travellers. Taormina, I am sorry to say, I never saw, and it is one
of my failures which I most regret; but I have
a vivid impression of the extraordinary beauty of
Palermo and all its surroundings; I confess I should

be inclined to prefer it even to Naples, though I admit that this may well allow of dispute. The last time I was at Cambridge, I remember, I saw in the Fitzwilliam museum a large picture of Taormina, which certainly gave me the idea of a scene of wonderful beauty.

(December 18th). I was delighted when I saw your handwriting this afternoon, and am very happy that you are safe in London and have not caught any illness on your journey. We were really beginning to be nervous about you. I would have written to you to Paris, but I did not know your address there, and indeed did not know that you meant to make any stay there.

The weather has now changed, I am glad to say, and to-day has been very fine, though with a coldish wind.

We are quite alone but expect (with some uncertainty) a flying visit from Harry on his way to Marchfield. I do not know whether you have heard that Willie has lately arrived there from India, and is staying there with his bride, with whom both Cissy and Emmie are delighted.

Fanny is tolerably well, and sends you her best love.

Ever your very affectionate brother,
CHARLES J. F. BUNBURY.

1882.

JOURNAL.

1882.   Yesterday I received the first proof-sheet of a book which Spottiswoodes are now printing for me, for private distribution. It is to be called "Botanical Fragments," and is a collection, partly of reprints (with corrections and additions) of papers published in the Linnean Transactions and elsewhere, partly of unpublished and incomplete essays on botanical subjects. I first wrote to Spottiswoodes about it on November 20th, and sent them the first portion of my MS., but from inexperience, I did not send enough for a complete printed sheet of 16 pages; and thus the beginning has been delayed, so that I did not receive the first proof till yesterday. This is much more correctly printed than I expected.

———

Bitter weather :—the winter has begun very early, and with great severity. The first snow fell on the 7th : *here*, there was not a great fall, and no remarkable wind; but in many parts of the country, especially in Wales and Scotland, as appears by the newspapers, there were terrible snow-storms—roads and railways blocked, trains buried, and some lives lost. Ever since, the ground has continued covered with snow, and the frost severe, except on the 13th, when there was a partial but incomplete thaw.

———

December 18th. 1882.

A talk with Lady Hoste—very cheerful and good-humoured, as indeed she always is.

Had a letter from Edward announcing his safe arrival in London—a great comfort. The last before this, received on the 8th, was written from Messina, and gave an interesting account of his tour in Sicily, which had been very prosperous; but we had not heard of his safe arrival at Marseilles, and I was moreover afraid he might have suffered from the severity on the journey through France. He is 71 years old.

————

December 19th.

Fanny had a charming letter from dear Rose Kingsley, of whom we had scarcely heard for a long time. She says that her own health has been very much restored by her stay at Spa in the autumn; it seems to have given her quite a new life. She is exceedingly busy, writing for various American periodicals. Her mother is now quite confined to one floor of the house—unable to go up or down stairs without danger of a serious attack; but otherwise tolerably well for the time.

========

LETTER.

Barton,
December 23rd, 1882.

My dear Susan,

I begin a letter to you in the hope that it may be finished in time enough to reach you on

1882. your birthday.   I was extremely glad to hear of
Katharine's, Rosamond's and Dora's safe arrival at
Florence, and you are now a large and lively and
happy party assembled together—four of the elder
generation and four of the younger, and I hope and
trust that nothing will happen to cast a cloud over
the brightness of your Christmas and New Year.
We have been (till yesterday) unusually quiet—
absolutely alone for rather more than a month.
Just now, indeed, Harry is with us, but he only
came yesterday and goes off to-morrow to Marchfield, where his brother William is with his mother
and sister and his newly-married wife.   Harry
however, I hope, will come back to us for a longer
visit before he returns to Antwerp.

You will have heard by Fanny's letters how many
deaths of friends—or of friends' children—we have
had to mourn within a short time; above all, those
of our dear Edward Campbell and of the good
Archbishop Tait.   This last is a grievous public as
well as private loss, but he was one who might well
say—to die is gain.

I have lately been reading aloud to Fanny in the
evenings some parts of my old journals of 1867-69;
the refreshing thus one's memory of past time
produces a curiously mingled feeling of pain and
pleasure—and yet it is hardly quite pain, either, but
a sort of gentle, modified sadness—"there is a
pleasure in this pain."   The same feeling is produced by your father's letters, of which Fanny
reads a great many to me, and which are full of
memories of departed friends.   He wrote admirably

good letters. Fanny is indefatigable in working at 1882.
that book, but it will be long before she can begin
to print.*

She has lately read through a bulky volume, the
" Life of Henry Erskine," with which she is
delighted. I hope also to read it, but have as yet
read only the *Quarterly Review* article on it, which is
extremely entertaining. I am engaged on a still
huger volume, and that a *first* volume only, of the
"Life of Sir William Rowan Hamilton," (of Dublin,
not the famous metaphysician of Edinburgh).
He was an extraordinary man, but a more extraordinary child, if we are to believe what is gravely
asserted in this book, that at the age of *four years*
and a half, he knew Latin, Greek, and *Hebrew!*
John Stuart Mill was nothing to him! However, he
really was a man of extraordinary power and versatility of mind—was a first rate mathematician and
astronomer, a metaphysician also, and a poet, and
wrote very pretty verses, and aspired to be a great
poet; but Wordsworth showed him his mistake. His
correspondence with Wordsworth is very interesting,
so is that with Pamela Lady Campbell. To say the
truth, I have taken up this book mainly because its
editor Mr. Graves is a particular friend of my
cousin Emily Napier. I do think that it might be
improved by a good deal of pruning.

(*December 26th*). We have been quite alone this
Christmas, but there has been plenty of society and
merriment in the servants' hall and housekeeper's

* I could not continue it after my husband's death, so my sister, Mrs. Lyell,
undertook the Life, and did it admirably.—F. J. B.

1882. room, when I hope the servants and their friends
have had much enjoyment. They were a party of
45 at dinner yesterday. I hope you also (you in the
plural) have much enjoyment of Christmas, and I
hope you have had finer weather for it than we have
here. Christmas eve indeed was very fine, though
cold, but yesterday was a wretched day, wet, cold
and dark; and to-day though less wet and cold,
has been extraordinarily dark and dismal. I have
been pitying the poor holiday-makers very much.

Edward returned to England only a few days
ago, having been abroad ever since the beginning
of November. He first visited Corsica, and saw
something of its mountains and forests; and after
spending some time at Naples, made a tour in
Sicily, with which he was very much pleased. He
was especially delighted with the beauty of Palermo
and Taormina. Sicily seems to have made great
progress, especially in the way of roads and ac-
commodation for travellers, since I saw it 39 years
ago.

I will now close my letter, repeating my hearty
wishes for a happy new year to you and all your
family party, and as many more years as may be
most conducive to your happiness. Fanny had
caught cold yesterday, and coughed so much in the
night as to make me uneasy, but to-day she seems
to be pretty well. I am quite well except my deafness, which is worse than ever. Believe me,

Ever your loving brother,

CHARLES J. F. BUNBURY.

JOURNAL.

Mr. Harry Jones preached an admirable sermon 1882. in our Church this morning on the text: "Watch ye "therefore, and pray always."—(Luke, chap. 21). He dwelt on the necessity of watchfulness, especially in these times; that we should not be absorbed merely in the petty cares and immediate surround-ings of our daily life, but be observant also of those matters which might involve the greater interests of our country or of mankind; pointing out how the wonderful ease and rapidity of communication between the most distant countries, in these days, may easily deceive us as to the importance of passing events, hinder us from seeing them in their true proportions, and therefore from being prepared for their consequences. He spoke also very forcibly on the other clause of the text, the duty and neces-sity of prayer to render our watchfulness useful. His sermons are very powerful, excellent both in matter and language.

December 25th.

Mr. Harry Jones, in performing divine service this morning, omitted the Athanasian Creed, in which I think him perfectly right. In the sermon, speaking of Bethlehem, he gave us in a few striking words, his impression of that memorable place, as he saw it a few years ago.

December 29th.

We are now again very near the close of one year

1882. and the beginning of another, and I feel it both a duty and a pleasure to express my deep and humble gratitude to Almighty God for all His goodness to me; for the many and great and undeserved blessings which I am permitted to enjoy. Above all, I can never be thankful enough for the blessing granted to me in a wife whose merits can hardly be over-praised, and whose devoted love to me is the constant source of my happiness. Neither can I sufficiently express my gratitude for the good health which has long been granted to us, and which at my age (nearly 74) may be considered an uncommon blessing. I am not indeed strong (but as to muscular strength, it is what I never did possess in any considerable degree); but I am very free from any painful or harassing or disabling ailment; my mind is (I hope) clear and sound, and my eyes (an especial blessing) quite serviceable.

A sadness has been cast over this year by the death of numerous friends. Very early in the year, in February, our dear, good old Janet Rennie departed, as I have recorded in my journal. For several months afterwards there were no deaths which particularly touched us; but from the 25th of November, when we heard of the death of our very dear friend, Edward Campbell, they have come very fast, as I have noted on the 6th of December. And even since, another name is to be added to the list —that of Colonel Anstruther; a very worthy man and a thorough gentleman, but interesting to us chiefly as the brother-in-law of our dear Mrs. Mills. Our contemporaries are passing very fast off the

stage, as must indeed be expected at our time of 1882.
life, and the departure of each of them is a fresh
and useful warning to us to put our house in order;
but when the young are taken away, those on whom
seemed to rest the hopes of the coming age, and to
whom we have trusted to cheer and enliven us in
our declining days —*that* is a more severe pang.
Happily this worse affliction has not visited us this
year.

Another friend who has departed in the course
of this year was Mr. Donne (William Bodham
Donne): a very accomplished and remarkably
agreeable man. He lived several years at Bury,
having settled there at first for the advantage it
offered for the education of his children; and when
we lived at Mildenhall we saw him frequently; he
came often to stay a few days with us, and more
than once gave lectures at the Institute. After he
removed to London we met less often, and less so
latterly, though without any estrangement or loss of
friendly feeling.

The two most celebrated men who have died in
this year, I think, were Charles Darwin and Garibaldi; celebrated for very different achievements.
With Darwin I was acquainted a long time (ever
since the year '42) but never very intimate, though
I liked him much. Garibaldi I never saw but at a
distance. But certainly, of all the list, the one who
must be considered the greatest public loss, was
Archbishop Tait.

I ought certainly to add Longfellow, the great
American poet, to the number of the most famous

1882. men who have died in this year. He was, I should
think, almost, if not quite, as much read and
admired in England as in America. Other men of
note who have died in the year, with whom we were
not personally acquainted are :—

Sir Charles Wyvill Thomson ; Sir William
Palliser ; Sir Woodbine Parish ; Sir Henry Cole ;
Dr. John Brown, the Author of " Horæ Subsecivæ."
Sir George Grey ; Dr. Pusey ; Dr. Hawkins (of
Oriel).

I ought not to omit, among our losses of friends,
that of the Duke of Grafton ; not that we were at
all intimate with him, but he was always very
friendly in his behaviour to us, and was a very
amiable and good man. I noted his death in this
journal on May 22nd.

Agreeable incidents in the course of this year have
been :—the marriages of William Bruce with Mimi
MacMurdo ; of my nephew William to Miss
Ramsay ; of Emily Egerton with Mr. Fielding ; and
of Jane Broke with Mr. Saumarez ; also (more
recently) the engagement of Lady Evelyn Finch-
Hatton to Mr. Upton.

Our visits in June, to the Charles Hoares at
Minley, and to Lady Grey at Cobham, were quite
delightful, combining the enjoyment of beautiful
country with that of the society of dear friends.

In the Cobham excursion I had moreover a
pleasure which I had not tasted for many years—
that of botanizing *in the field*, seeing several interesting plants in their native places of growth.—(See
this journal, July 3-4).

I had much enjoyment while in London, in my 1882. visits to the mineral and geological departments of the British Museum in South Kensington.

The establishment of Mr. Harry Jones as our clergyman in this parish, with frequent opportunities of conversation with him, has been a great satisfaction and a great source of pleasure to us. We find him (as I had hoped) not only a powerful preacher, but one who is likely to do much towards the moral and intellectual improvement of the parish, as well as a very agreeable and accomplished man.

I have always regretted the banishment of dear Lady Muriel Boyle to Ireland (see May 6th), and I am afraid we cannot look forward to her return within any definite time ; for Courtenay Boyle, although his work is very hard, finds it so interesting, that he would not willingly give it up even if he thought it right to leave Lord Spenser, and his wife would on no account leave *him*.

We have had very agreeable letters from her, but I am afraid her health is not good.

Of public affairs I will say as little as possible, because I can say nothing pleasant, and indeed I think of them as little as I can. The short campaign in Egypt, indeed, was well executed, and reflected lustre on our generals and our soldiers : and we may be very thankful that it was so soon over, for short as it was, it cost many valuable lives, and more by disease than by the sword. It is curious that Gladstone and his Cabinet, who came into power proclaiming, as their leading principles,

1882. Peace and Economy, should now put forward a triumphant war as their great success, and as the greatest title to the confidence of the country.

In Ireland there certainly does appear to be one symptom of improvement:—the intimidation of juries does seem to be in some degree baffled, and in several cases lately, murderers have been convicted: but I am much afraid that the improvement is only temporary, and would disappear if the Act for the repression of crime were not continued in force.

1883.

JOURNAL.

January 2nd.

1883. News of the death of Gambetta, causing a great sensation in France. He was only 44. I did not know enough about him to make any remarks on his character, or to foresee in the least what may be the consequence of his death. At the time of the Franco-German war, and for some time after, he was certainly a personage of importance, but since then, it has appeared difficult to follow or to understand his political cause.

January 5th.

The news of the death of General Chanzy, so famous in the war of 1870, followed—as it seemed—almost immediately on that of Gambetta.

I saw announced in *The Times*, the death of my old tutor Mr. Matthews (Frederick Hoskyns Matthews), at the age of 84.   He was a little more than 10 years older than I.   My father engaged him in 1822 as private tutor to my brothers and me, and he remained with us till we went abroad in the latter part of '27.   He knew thoroughly what he professed to teach — Latin and Greek and mathematics, and I think he was a very good teacher.   In classics, at least, I know that his instructions have remained very much impressed on my memory : so that whenever I read or recal to mind particular passages of the ancient authors, his particular modes of construing them—his favourite phrases, or his peculiar objections to others—constantly recur to my thoughts.   He was a good, quaint, simple-hearted man, a little hot and hasty in temper perhaps (I daresay we often gave him provocation), but very sensible to kindness ; he always expressed himself as very grateful to my father, and though I did not see him again for many years (28 years if I remember right), after we had parted in '27, he was very cordial, and expressed great joy at the meeting.   Since then, I have seen him a few times, as whenever he has been in London and has known of my being there at the same time, he has always come to see me, and shown the same cordiality.   So he has also to Edward, who has had more frequent opportunities of meeting him.   He was extremely gratified by the success of Edward's "History of Ancient

1883. Geography," and especially by its learning. Mr.
Matthews was well read in English literature, at
least in that of the 18th century, to which he was
very partial; indeed his taste was altogether
regulated by the canons of criticism which prevailed
in the time of our grandfathers; he was steadily
devoted to the "classics," in English as well as in
Greek and Latin. Of French, I think, he has but
an imperfect knowledge, and of other modern
languages none at all. Mr. Matthews was very fond
of billiards, and still more of chess, in which he was
a very great proficient.

He had many innocent little oddities and peculiar
ways, and was very shy and silent in society, had in
fact many of the characteristics of a collegiate
recluse. When I visited him at Hereford, in 1855,
at a time when he was the sole partner* in a
seemingly thriving country bank, and was comparatively a rich man, he received me in a very modest
lodging, and pointed out to me with pleasure and
pride, how like a room in college it was. When the
affairs of the bank turned out unfortunately, and he
was reduced to comparative poverty, I heard from
those who saw him often, that he did not merely
bear this reverse of fortune with equanimity, but
was actually pleased with it, and enjoyed his release
from all the cares and trammels of business and
wealth. The first time he went to call on Edward
in London, he wore a smoking cap instead of a hat,
and read a newspaper as he walked along the
streets.

* Rather an inaccurate phrase by the way.

Our little dinner party of eight—very agreeable ;
Mr. and Mrs. Harry Jones, Mr. and Mrs. Livingstone, Mrs. Storrs, my nephew Harry, and ourselves.
The two clergymen, though very unlike each other,
are both remarkably intelligent, cultivated, agreeable
men. Mrs. Storrs very lively and pleasing.

January 23rd.

My nephew Harry left us, after staying with us—
and most of the time alone with us—for nineteen
days. I am very much pleased with him. He is
thoroughly amiable and good, with an excellent
heart and sound principles, and very pleasing
manners. His chief fault is indolence, in which he
resembles me. Without any brilliant talents or
unusual power of mind, he has very fair abilities,
and if he concentrates his powers on any one
pursuit, he may, I think, make a very good figure.
His favourite pursuits hitherto have been chemistry
and landscape painting. The latter is the study
which he is now seriously pursuing at Antwerp, but
I have had no opportunity as yet of judging of his
progress in it.

January 31st.

Mrs. Storrs, the wife of one of the Bury clergymen, who has been staying with us since the 22nd,
left us to-day. Our acquaintance with her is not of
very long standing ; we first met her when we were
on our last visit to Lord Hanmer, in 1880, when

1883.   with him, we went to visit her father, Major Cust, at
Ellesmere.   She was at that time unmarried, but (if
I remember right), already engaged to Mr. Storrs.
Since they settled at Bury, my wife has kept up and
improved the acquaintance, till now it has become
a real warm friendship. Mrs. Storrs is indeed a
charming young woman, both attractive and estimable; with gay spirits, a lively and playful
imagination, great enjoyment of merriment and
drollery ;—at the same time, very earnest in serious
matters, full of kindness to the poor and to all who
need it,—and I have no doubt, an excellent wife,
and a valuable help to her husband in his profesional
pursuits. She is clever, well educated, well read,
and has thought on what she has read.   In short,
she is a valuable acquisition to our circle of neighbours and friends, and I heartily wish that she may
remain in it as long as we.

Another great favourite of mine, Mrs. Maitland
Wilson, was also with us from the 23rd to the 27th.

The weather throughout this month has been
to an extraordinary degree variable and unsettled ;
—hardly three days together the same.   No severity
of frost ; the lowest the thermometer has reached
was 23 degrees Farenheit ; and that only twice
in the month ; snow has fallen, I think, only once
(in the night of the 24th to the 25th) ; and in 13 or
14 days (I mean periods of 24 hours), the thermometer has not been down to freezing point.   But
there has been a great prevalence of damp, gloomy,
chill, sunless weather ; and almost alternating with
this, frequent storms of wind and rain.

The temperature yesterday morning at Bury 1883. (according to Dr. Macnab's information), was 20 degrees lower than at the same time on the 29th.

———

February 4th.

My 74th birthday. I return most humble and hearty thanks to God Almighty, for having permitted me to live to this age in such comparatively good health and in the enjoyment of so much comfort and happiness. I can only repeat what I wrote on the 29th of December of my deep sense of gratitude for the many blessings granted to me, and above all for my union with my invaluable wife.

Just at present, we are both suffering some little discomfort; from having caught colds, but this evil, I trust, will not last long.

═══

LETTER.

Barton Hall, Bury St. Edmund's,
February 5th, 1883.

My dear Katharine,

Very many thanks for your kind letter and good wishes for my birthday ; it was not so cheerful as some I have known, because we were both confined to the house with bad colds, and had to spend much of our time in coughing, &c. To-day, I hope, we are both better ; I certainly am, though confined for the present to one floor. At any rate, though I am not perfectly well, I have great reason to be

1883. thankful for being as well as I am, and free from any painful or serious ailment at the age of 74.

I can well understand what pleasure you must have in going about with Charlie at the Zoo; he is a most interesting child, and I should very much like to see him again.

My *book* is now in print as far as page 144, which I calculate will be about one-third of the way through. The printing has been very accurate, so that I have had little trouble in correcting; indeed I have been quite surprised to find how correctly the scientific names have been printed. But a book does not look pretty " *en deshabille,*" that is, in proof sheets.

I see in *Nature* that our old friend (Fanny's and mine, I mean), the Peak of Teneriffe, is infected by the spirit of this age of *agitation*, and wants to have a *pronunciamiento* of its own. I wish I were " there to see," as John Gilpin says. I remember vividly and so does Fanny, the fragment of the town of Garachico, which the Peak demolished in its last outbreak, and the current of lava remaining on the precipitous mountain side, like a cataract, turned into stone.

I do not know how lately Fanny wrote to you, but I dare say she has told you all about the pleasant time we had in the latter part of last month, when Mrs. Storrs was staying with us. She is really charming, lively, clever, amusing, very well educated and well read, with excellent feelings and principles; a great acquisition to the neighbourhood. I wish she may not be taken away from us, by her husband being

selected to succeed Mr. Wilkinson at St. Peter's,— 1883.
for which he is talked of. We are very fortunate in
having two such clergymen as Mr. Harry Jones
here, and Mr. Livingstone at Mildenhall, we like
each of them more and more as we know them
better.

With much love from both of us to Rosamond and
yourself,

Believe me ever,

Your loving brother,

CHARLES J. F. BUNBURY.

Barton,

February 8th, 1883.

My dear Susan,

I thank you heartily for your very interesting as well as very kind letter which I received this
morning ; it was not at all the worse for coming a
little too late for the actual anniversary of my
birthday, for I have thorough confidence in your
affection, and feel sure that you would think of me
kindly, not on my birthday only, but everyday. I
am very much interested by your remarks on the
books you have been reading, though perhaps most
of them are little or not at all known to me. Of
Whittier I know only two poems, but should be
glad to know more : one of them is that which
begins—

" I mourn no more my vanished years."

It is the last poem in Trench's collection : I think
it very beautiful : I have read it over again since I
received your letter, and think there is a great deal

1883. of gentle wisdom and mild pathos in it. The other piece of his that I have read, is a little story called (I think) *Maude Müller*, which is very pretty. I do not think I ever read Dennis's "Etruria:" at any rate if I did, it was so long ago that I have quite forgotten it. I have not heard of Hamerton's "Life of Turner;" I suppose he is the same Hamerton who has written a book (or books) on "Country Life in France:" so far, I believe, he is good; is he likely to be a good authority on painting?—I do not know. I have never read "*Les Misérables*," and do not wish to read it, as I have no wish to be made miserable, and I have been told that the effect is quite in accordance with the title of the book. Besides, I dislike Communists and their writings.

I am very glad you are going to read Froude's "History of England." It is beautifully written, and as interesting as any romance; indeed I do not think Walter Scott himself ever wrote a more brilliant narrative than Froude's — that of the "Rising in the North" in Henry the Eighth's reign. Of course you will read him (Froude) with caution and ample allowance, remembering that he is *always*, on every question, an advocate, and not a judge. It is very interesting to compare what he has written about Mary Stuart with what Burton—in his "History of Scotland"—says on the same subject.

I have just finished reading a very pretty historical novel or romance, called the Burgomaster's Wife, translated from German; I daresay you will have

heard of it from Fanny, as she recommended it to 1883.
me. It is remarkably well written.

I have lately read Henry Erskine, with the
review* of which (in *The Quarterly*), I had been
delighted: but I have found the book itself much
too long; the reviewer has picked out pretty nearly
all that is entertaining, except perhaps the Scotch
people.

Pray give my love to Joanna and Leonora, and
very many thanks for their kind wishes. Tell
Joanna I have just got Geikie's new " Text-book of
Geology," which is about the thickest octavo I have
seen—above 900 pages: but I have no intention of
reading it *through*.

Again thanking you for all your kind words and
good wishes, I am ever,

Your loving brother,

CHARLES J. F. BUNBURY.

JOURNAL.

February 10th.

Since the 5th, I have not been able to go out of
doors except yesterday, when the weather was very
fine; I took a half-hour's drive with Fanny. But I
have plenty of occupation in doors.

Edward wrote to me from London on the 4th:

" Many people are alarmed at the prospect of
" things in France, but I hear that well-informed
" Frenchmen have no apprehension of any immediate
" outbreak or disturbance, though they do not pro-

* By Lord Moncrieff.

1883. "fess to see their way to the future. But it is
"evident that Gambetta was the one master spirit
"that has kept things quiet until now, and the way
"that all the elements of confusion have broken out
"as soon as he was taken away is the greatest
"possible tribute to his memory."

We were told some little time ago, that the living
of Clovelly had been offered to Mr. Harrison, Mrs.
Kingsley's son-in-law, and more lately, that he has
accepted it. Fanny thereupon wrote to Rose,
saying that she did not know whether to congratulate or condole with her. Yesterday she had a
very interesting letter from Rose, saying "cer-
"tainly congratulate us," and to-day she had
another from Mrs. Kingsley to the same effect.
They seem delighted, that Charles Kingsley's
daughter should be settled in a place so dear
and "almost sacred" to them all. Clovelly,
indeed, as I knew, is not only an exquisite place
in itself, but one for which Kingsley had an
especial love; and besides, its climate will be of
the greatest advantage to Mr. Harrison's health,
which has suffered from what Rose calls the "arctic
climate" of Wormleighton.

Of course they feel that the removal of the
Harrisons to such a distance will be a very great
loss to them, but they feel also that it *must* have
come sooner or later, on account of Mr. Harrison's

health : and that on the whole the advantages 1883.
preponderate.

Clovelly is a poor living. It has been (Rose says)
shamefully neglected for many years, and therefore
a great part of the population are Dissenters : but
they are a noble race of men.

Rose speaks of the mischief done by the incessant rains, and says—"we are being slowly
drowned."

February 15th.

The accounts which come from Ireland—of the
judicial inquiry now going on at Dublin, and the
gradual detection of the organized system of murder
which has been so long in operation, and specially
its working in the instance of the Phœnix-park
murders—are painfully, terribly interesting. It may
now be reasonably hoped that the actual operators
in those horrors may be brought to justice : and if
so, I hope no mercy will be shown to them ; but
whether it will ever be possible to discover and to
crush the *greatest* criminals—the original contrivers
and supporters of the whole infernal system—is, I
fear, much more uncertain ; I cannot believe that
the low, vulgar ruffians who are at present in
custody, can have been the authors and contrivers
of such an elaborate design, or could have had
the means of carrying it on.

February 19th.

Scott went to Mildenhall, 3 days ago, to inspect

1883. the state of the Fen, as it was feared that the almost continual wet weather lately might have caused an inundation.   His report is very satisfactory; he says that he has seldom seen Mildenhall Fen in a better state at this time of the year.   The level of the water in the river is indeed high, but the action of the steam engine and the constant watchfulness of the men in charge of it have been so efficacious, that the bank has nowhere given way in the least.   There is no inundation, nor any present danger of it; there is little depth of water in the ditches, and the surface of the land is dry enough for horses to be at work in ploughing it. In fact, he says, the land in the Fen is in a better state than about Bury.

———

February 22nd.

The excessive wet of these last 3 months has not only been dreary and depressing in its physical effect on ourselves, but makes one very anxious and uneasy as to its probable effect on the land and on the prospect of the next harvest.   I speak of the excessive wet; but in fact, it appears, the actual rainfall has not been very much above the average : but the number of rainy days has been uncommonly great, the amount of sunshine uncommonly small : thus the ground has had no sufficient time to dry, it has been kept continually soaked and sodden, and not at all in a fit state for the operations required by the time of year.   The farmers' prospects are very gloomy.   We must only try to have a firm faith in God's goodness.

Our dear old friend Mrs. Rickards died this day.
She was nearly 86 years old, and had for some time
past been bed-ridden and helpless, entirely dependent on others,—indeed scarcely more than half
alive, so that her best friends could not wish her
earthly existence (life it could not be called) to be
prolonged. The news of her departure (long
expected) was brought to us by Mr. Reid, the
clergyman of Stowlangtoft; he told us that she
passed away quite calmly and peacefully, without
the least struggle, and that the expression of her
features after death was beautifully serene. It is a
comfort to know this. She has had every comfort
in her long decay that her condition could allow
of:—my wife and Mrs. Maitland Wilson have
alternately visited her almost every day for several
weeks past, and have sent her everything that they
could think of as conducive to her comfort; she has
had some good friends also in Bury, and her own
maid, Mrs. Capon, an excellent woman, has been
most thoroughly devoted to her.

March 1st.

My acqauintance—indeed my friendship—with
Mrs. Rickards and her husband, began more than
forty years ago; I am not certain of the year, but
it was in 1840 or 1841:—I used often to walk over
to Stowlangtoft from this house to enjoy a chat with
them, taking great delight in their vast variety of
knowledge and their simple, easy manners. After I

1883. married, Fanny soon grew as fond of them as I was, and though, owing to the distance of Mildenhall we could not see them as often as I had been used to do, the acquaintance was not suffered to drop. In 1852 we saw a good deal of them at Ventnor, whither they had gone for their daughter's health and I for my wife's : and after we came to live here, the old intimacy was actively kept up till the death of Mr. Rickards, which was a grievous loss to us and to the whole neighbourhood. It is a great pleasure to me to look back on the many conversations I had with them, and especially with him, in that pretty parsonage of Stowlangtoft,—conversations on a vast variety of subjects, for on almost any subject he had something to say, and something worth hearing.

March 10th.

A severe and very unseasonable winter. January was, as I have already noted, generally mild, with scarcely any snow, and no severe frost, but a great deal of dull, damp, gloomy weather. February still more exempt from frost and snow, with a larger proportion of fine weather ; towards the latter part of the month especially, several fine days. In this month, from the 2nd to the 5th (both inclusive) the weather was beautiful. But on the 6th came a tremendous N.W. gale, with frequent and furious snow-storms. The 7th was still worse—snow-storms in rapid succession all day, with much wind, and the snow lying at nightfall. Since, the wind has been moderate, and the sunshine bright and clear, 1883. so that in places sheltered from wind and exposed to the sun, it was even warm. But the snow has lain, almost undiminished (except in very sunny spots) and there has been fresh falls, though at long intervals. The frost very sharp at night: down to 18 degrees Fahrenheit, in the night of the 7th-8th; to 17 degrees in that of the 8th-9th—and to 14 degrees last night.

To return to Mr. and Mrs. Rickards. I do not mean that their knowledge or their conversation were at all confined to natural history,—*he* in particular, had a great variety, a great range of knowledge; and I had great pleasure in hearing him talk (as he would freely and very well, at the same time with the greatest simplicity) of the classics ancient and modern, Latin and English. I have no doubt he was a good Greek scholar, but on that ground I could not meet him. I liked also to get him to talk of his old days at Oriel, and the famous men he had known there; for he had lived in intimacy with that remarkable set—Newman, Hurrell, Froude, Keble (I think).

————

March 17th.

On the 5th, Sir Henry Wilmot (Mrs. Rickards's nephew) and Mr. Francis Wilmot came to stay with us in order that they might attend Mrs. Rickards' funeral at Stowlangtoft; it took place on the 6th (the first day of the snow), and they stayed to the 7th.

1883.   Sir Henry Wilmot is an accomplished man ; talks much and well ; has seen much service (he gained the Victoria Cross at Lucknow) and has " seen the " cities and observed the characters of many men." He spoke highly of Sir Garnet Wolseley, whom I have observed to be, in general, by no means a favourite with officers of the army ; Sir Henry said he was a remarkably well-read soldier, devoted to his profession, indefatigable in the study of it, and thoroughly well acquainted with the theory as well as the practice.

March 6th, we had a very pleasant little dinner party—Mr. and Mrs. Bland, Mrs. Storrs, Sir Henry and Mr. Wilmot.

The 12th.   Darling Sarah arrived—to our great delight.   Also John Herbert.

Yesterday we had the very startling news of the terrible dynamite explosion at the Government Offices at Westminster.   If it was not accidental, (and almost everyone seems to believe that it was *not*) it is a horrible proof of the excess of wickedness which is lurking around us, and of the amount of mischief which might be done without risk to the perpetrators.

It is surprising, and a very great mercy, that on this occasion no one was hurt, but we cannot hope that future crimes will be equally unsuccessful.

—————

March 19th.

The Bishop of Ely* has just left us; he was coming into the neighbourhood to hold confirma-

* Dr. Woodford.

tions; we invited him to this house, and he came to 1883.
us with Archdeacon Chapman, Mrs. Chapman, and
their son and daughters, on Saturday, the 17th.
We have had a large party to meet him; besides
Minnie Napier, Sarah Seymour, John Herbert
and Dora Pertz, who have been with us longer,
we have had Lady Rayleigh, Richard Strutt,
Colonel Corry, Mrs. Wilson and her very pretty
daughter Ida, staying in the house; and the Storrs,
the John Paleys, Major and Mrs. Harris, and two
other officers, dined here on the 17th.

It has been very pleasant; the Bishop is a very
agreeable man, with peculiarly mild, insinuating,
winning manners, and a great deal of good conversation and anecdote. The Archdeacon, a very
agreeable man, and so also is Colonel Corry.

March 21st.

A thaw began on Monday the 19th, and by the
morning of the 20th, the snow was nearly all gone;
no more has since fallen, but the weather is still
very cold.

March 23rd.

Dear Sarah Seymour left us on the 20th, and her
dear mother on the 22nd; I was very sorry to part
with both.

Our good neighbour, Admiral Horton has died
almost suddenly. He was in London last week,
and on the 14th or 15th, presented his son Sydney
at the Levee. He returned home on Monday last,
the 19th; began to feel unwell the same day, soon

1883. became unconscious and continued in the same
state of insensibility till yesterday evening, when
he died, soon after 9 o'clock, without recovering
consciousness ; happily also, we may believe, without suffering.

I am very sorry for his wife, who was devoted
to him ; also for his son, who, at the age of just one-
and twenty, is thus launched on the world without
the protection and advice of a father.

The Admiral is a loss also to the neighbourhood,
and to us, for he was always a courteous, kind and
friendly neighbour. The Admiral was a very active
and useful magistrate, a clear-headed and energetic
man, who took a considerable part in the business
of the district ; He was a man of strong religious
principles, a resolute and active supporter of what
are called Low Church or Evangelical opinions.

April 14th.

Here is a great gap. I must go back.

The day before I noted in this journal the death
of Admiral Horton, our dear Bruces—Willie and
Mimi—arrived ; and their society, which lasted till
the 9th of this month, was a very great pleasure to
me. They are both delightful. It is seldom indeed
that one meets with a young couple, *both* so richly
gifted with all sorts of good qualities, personal,
intellectual, and moral. Both very handsome. It
is delightful to see how much attached they are,
and how happy together. God grant them long life
and health, and that they may long have the

power (as they certainly will have the wish) of 1883.
dispensing happiness to a wide circle around
them.

During the stay of the Bruces, we had also from
time to time some other detachments of pleasant
guests.   Katharine, Rosamond and Dora stayed
but a short time ; since then we have had Bernard
Mallet (Louis Mallets eldest son, a very pleasing
and interesting young man), the Livingstones from
Mildenhall, and, especially, Mr. and Mrs. John
Gladstone. *She* was a comparatively old acquaintance, Mrs. Laurie's niece, formerly Miss Constance
Bayley, but Mr. John Gladstone is a new acquisition to our circle, and a very welcome one,
being a remarkably agreeable man, cheerful and
lively, and very well informed.   He is particularly
fond of trees, which subject served to draw us
well together.

LETTERS.

Barton,
April 16th, 1883.

My dear Leonora,

Fanny tells me (and I am afraid she is in
the right) that I have never written to you since
receiving from you a very kind letter of congratulation on my birthday.   It is very wrong, and
I acknowledge my fault and promise never to do so
again—till next time.   In truth I am a very bad
correspondent, and my only excuse is, that my hand

1883. is so cramped, it is more labour for me to write one letter than for you or one of your sisters to write six; you may have heard from Fanny that I have been rather out of condition for some time, having caught the influenza—or rather been caught by it—about 3 weeks ago: and though it never was very severe, it has hung about me rather obstinately, and I am only now shaking it off. Unluckily it spoiled my enjoyment of Katharine's too-short visit, and prevented my seeing much of Dora, with whom I would gladly have had more confabs about botany. I wish she and you too could be here now, to see the very beautiful Orchids which are in blossom in the hothouse:—Dendrobium nobile, D. fimbriatum, Odontoglossum Roezlii, Sobralia macrantha—all gloriously beautiful.

I suppose the wild flowers in Florence are now coming into beauty, though I fancy the season has been very bad and backward, as we hear it has on the Riviera; here, the flowers in general, as well the wild as the hardy garden ones, are from 10 to 14 days later than last year.

I am getting on with my "Botanical Fragments," of which I suppose you will have heard from Fanny or from Katharine. Fanny goes on reading to me, as she proceeds, her collections for the Life of your father,* and what she has last read to me has been particularly interesting to me, being his letters and journals during the last visit to Italy. They bring back most vividly to me, beautiful Florence

---

* This I gave up writing after my husband's death, to my sister Mrs Lyell, to complete.—F. J. B.

and all its charms and objects of interest. Fanny's 1883. industry is really wonderful.

We have lately enjoyed a most pleasant visit from the Bruces, William Bruce and his wife, who you know was Mimi MacMurdo. We have lately also had another very pleasant couple staying with us,—Mr. and Mrs. John *Gladstone*, he is a second cousin, I understand, of the Prime Minister, but of very different opinions.

With my best love to Susan and Joanna, and love to all your party.

Ever your loving brother,

CHARLES J. F. BUNBURY.

———

Barton Hall, Bury St. Edmund's,
April 17th, 1883.

My dear Katharine,

I thank you very much for the photograph of Joseph Hooker which you have so kindly sent me. It is, as you say, an admirable photograph, and the best likeness I have seen of him as he is now; I value it very much, and am grateful for your kind thought of me in asking for it.

I have been much interested by your account of poor old Mr. Bentham, and feel very sorry for his loneliness. As you say, it is a great pity he does not accept the Hookers' kind offer. It is a wonderful thing that, at his age, he should have such energy and such an unclouded intellect as to have been able to carry through to the end his part of the Genera Plantarum; but even when the energy and intellect are such as *his*, it is a melancholy thing to be left so alone in the world—a

1883. melancholy thing to outlive all whom one cares for; and I think our friend Admiral Horton, who was snatched away so suddenly, without any lingering decay, from the midst of those he loved, was not at all to be pitied, but rather to be considered fortunate in his fate.

I was very unlucky in losing so much of the enjoyment of your visit through the troublesome influenza. I hope it has at last pretty well taken its departure, but it has been a very tenacious *bore*.

I wish you could see our Orchids now: we have such beauties; the delicate intricate filigree-work fringe of the lip of Denbrobium fimbriatum is perfectly exquisite, and the whole flower is of the colour of a ripe apricot. And the Sobralia is magnificent; without flower one would take it for a reed rather than an Orchid, but the flower is of the richest possible purple. Odontoglossum Roezlii, and Dendrobium Wardianum, which I think you saw here, are also most beautiful: and all these, and some others, we have had now in succession for more than a month. Out-of-doors, I find that the wild flowers, and most of the hardy garden ones, from 10 to 15 days later in blossoming than last year: but they are almost exactly the same this year as in 1881.

You will have heard from Fanny all about our very agreeable visits from the William Bruces and the John Gladstones.

I am very glad that your daughter-in-law and your new little grand-daughter are going on well. I hope

Leonard will very soon go to a warmer climate, 1883. which must surely be of great importance to his health.

Believe me ever,

Your loving brother,

CHARLES J. F. BUNBURY.

<hr/>

JOURNAL.

April 18th.

News of the death of Lord Talbot de Malahide, at Madeira.

April 20th.

My nephew George arrived.

April 22nd.

We (George with us), went to morning Church— an excellent sermon from Mr. Jones.

April 23rd.

George went away.

April 29th.

The news of Mr. Storr's nomination to S. Peter's, Eaton Square.

May 2nd.

Arthur arrived on sick leave.   Fanny very happy and I very glad.   He went to dine at Livermere.

———

May 3rd.

A sore vexation.   Arthur ordered (though lame) to rejoin his regiment by 1st train ; he set off immediately.

———

May 4th.

Telegram of Arthur's safe arrival at Dublin. News of the defeat of the Affirmation Bill.

———

May 8th.

Poor old Mrs. Eagle (aged 87) came to luncheon.

———

May 10th.

The Harry Bruces and Mrs. Storrs arrived.

———

May 11th.

Morning fine.

I lounged in the garden with Fanny and the Bruces, for nearly an hour.

Mr. and Mrs. Heathcote, the Morewoods, the Montgomeries, and John Hervey arrived.

———

May 12th.

Dear Mrs. Storrs went away, much to our sorrow. Sir Brampton Gurdon arrived.   Mr. and Mrs. Saumarez, and the Blands dined with us.

May 13th.

A most beautiful day ; perfect summer. Was much in the garden with the ladies.

————

May 14th.

A mild but sunless day. The festival of the Barton Parish Club. Fanny and most of the ladies went to see the festivities at the Vicarage.

Mr. and Mrs. Livingstone arrived.

————

May 15th.

A most beautiful day, perfect summer. All our guests went away, except the Livingstones. Dear Katharine arrived. The Storrs, the Harry Joneses, and Mr. and Mrs. Algernon Bevan dined with us.

————

May 16th.

Another equally splendid day. The Livingstones went away. Fanny and Katharine went to the wedding of their cousin, Miss Anderson.

I lounged in the arboretum in the morning ; in the afternoon Fanny drove both of us in the pony carriage.

————

May 17th.

Again a splendid day. Fanny went to Mildenhall, and returned to dinner.

Strolled with Katharine in the arboretum and groves, in morning and afternoon.

————

Dear Katharine went away.

We went to morning Church and received the Communion. First time I have been able to go to Church since April 22nd; an excellent sermon from Mr. H. Jones.

We inspected the Ferns in greenhouse.

A most beautiful day,—perfect summer.

We drove out in the landau to Livermere; saw poor Mrs. Horton, and had a pleasant talk with Jane Saumarez and Freda Lorraine, who showed us the garden.

We drove out. Fanny and I visited poor old Patrick Blake, the Admiral.

A most beautiful day.

A pleasant visit to the Harry Joneses, and a delightful drive through the lanes full of hawthorne.

Fine, but more unsettled. Preparing for London.

Beautiful weather. The 39th anniversary of our happy marriage.

[During the months of April and May, Sir Charles was reading :—
Seeley's " Life of Stein."
Nasmyth's " Autobiography."
Geikie's " Geology."
Green's " History of the English People."
Dean Bradley's " Life of Arthur Stanley," and
" The Life of Sir Henry Lawrence."
We had visits from Mr. and Mrs. Saumarez and Sidney Horton and Admiral Blake.—F.J.B.]

June 1st.

Beautiful weather. We drove in the pony-carriage to the Shrub, and rambled there with much enjoyment, though the bluebells were fading.

June 2nd.

A splendid day, quite hot. Up to London. Arrived safe, thank God. Visits from Charlotte and Wilhelmina Legge, and Lady Arthur Hervey. Edward and Harry dined with us.

June 3rd.

Read prayers with Fanny.
We spent the afternoon very pleasantly with the MacMurdos at Rose Bank.—Mimi and Willie with them. Splendid weather.

June 5th.

We called on Mrs. Young, and afterwards drank tea with the Lady Legges, who were very pleasant; met there the Arthur Herveys and Mrs. Lyndoch Gardiner. The MacMurdos and Edward dined with us.

June 6th.

Fanny took me to the Natural History (British) Museum, where I spent an hour pleasantly.—Afterwards we called on Norah Aberdare. Mr. and Mrs. Harry Jones dined with us.

June 7th.

Foggy, gloomy and very cold. We dined with the Morewoods—met the Arthur Herveys and others. William Napier and General Ives came to luncheon with us.

Lady Colebroke to tea.

June 8th.

Pamela Miles came to luncheon with us. The Clements Markhams visited us, also Mrs. Ford.

June 9th.

Received and corrected the last sheet of my book.

My nephew Willie and his wife came to luncheon, also Mrs. Heathcote and Sir George and Lady Wilmot Horton. We went through part of the National Portrait Gallery—very interesting.

June 10th.

We went to morning Church at Eaton Chapel.

Very agreeable visits from Mr. Sanford and his daughter Ethel, and Alexander Kinglake. Edward and Clement dined with us.

June 11th.

Lady Grey, Mrs. Montgomerie and Ada Ridley came to luncheon.

June 12th.

A very fine and mild day. Dear Minnie arrived looking very well, to our great joy.

We spent an hour in the Natural History Museum—very interesting.—Afterwards visited Mr. Bentham. Kate Ambrose came to luncheon with us.

June 13th.

A very agreeable visit from the Bishop of Bath and Wells. Rosamond and Arthur Lyell, and Miss Chambers,* dined with us.

June 14th.

The Clements Markhams came to luncheon.

Our dinner party:—The MacMurdos, Mr. and Mrs. John Gladstone, Mrs. Laurie, Frederick Jeune, two Wilson girls, Bernard Mallet, Minnie, Harry.

June 16th.

Dear Rose Kingsley came to luncheon—her visits are a great pleasure to me. John Herbert dined with us.

* She became the wife of Arthur Lyell.

Cissy and Augusta came to luncheon.

Our dinner party:—Lord Coleridge, the Arthur
Herveys, Lady Grey, the Louis Mallets, Lady Mary
Egerton, Willie Bruce, the Leckys, Rev. Mr.
White, the Montgomeries and Harry.

June 19th.

We dined with Mr. Sanford and his daughter.
Met Arthur Hervey and his daughter Caroline
(Truey), also Mr. and Mrs. Moysey, Mr. Murray,
Mr. Ouless, the artist, etc.

Dear Minnie left us and went to her own house.

June 20th.

Our dinner party:—Leopold and Mary Powys
and a daughter, the Walronds and a daughter,
Admiral Spencer, William Napier, Minnie, Harry,
Clement and Mr. and Mrs. Ridley.

June 21st.

The Dowager Lady Rayleigh came to luncheon,
also Mary Lyell and dear little Charlie.

June 22nd.

We went to the Royal Academy, but could see
little for the crowd.—Then to Grosvenor Gallery;
where there were some good things.

We dined with the Douglas Galtons.

June 23rd.

Dear Mrs. Storrs came to stay a few days with us.
Dear Joanna came to luncheon.

We drove through the Park to see the rhodo- 1883.
dendrons.

Our dinner party:—Mrs. Storrs, Mrs. Laurie,
Minnie, John Herbert and Edward—very pleasant.

June 24th.

Mrs. Storrs staying with us. A long and very
interesting talk with Louis Mallet.

Read the 1st sermon of Allison, and 2 of Stanley's
on the illness of the Prince of Wales.

June 25th.

We went out to Fulham to dine with Miss
Sulivan :—met the MacMurdos, Professor Allmann,
Mr. and Mrs. Hugh Smith, Mr. and Mrs. Dyer, &c.

June 26th.

We dined with the MacMurdos at Fulham :—met
Sir Arthur and Lady Cunningham, Mr. and Mrs.
Moulton, the Leopold Powyses, Willie Bruce,
Edward, &c.

June 27th.

Down to Barton. Susan Horner and Colin
Campbell with us: arrived safe at home—thank
God.

June 28th.

Day fine and warm and pleasant, and the meeting
(in my grounds) of the Bury Horticultural Society

1883. was very successful—a good number of people and well kept up.   Many beautiful plants.

_____

June 29th.

Barton.

A most splendid day, very hot.   Spent most of the day strolling about the garden and arboretum, etc.   Not much mischief done by the horti-culturalists.

_____

June 30th.

Another beautiful and hot day.   I very unwilling to leave Barton; we returned to London: arrived safe—thank God.

_____

July 2nd.

Susan and Leonora came to luncheon.   A visit from Sir Arthur Cunningham.

_____

July 3rd.

A hot day, with a few showers.

We drove out—called on Katharine, saw her, Leonora, Joanna, Rosamond, Annie, Dora, Miss Chambers, Miss Zileri and Miss Nicholson—a pleasing bevy.

Minnie dined with us.

_____

July 5th.

Very fine weather.

Our   dinner   party:—Hookers,   Mr. and Mrs. Robert   Marsham,   the   Clements   Markhams,   the

Harry Joneses, Montague MacMurdo, Mr. Yeatman, 1883.
Guy Campbell, Miss Egerton, Minnie and Edward.

—————

July 6th.

We went by road to Kew, to Lady Hooker's
garden party: — met Miss North, Katharine,
Leonora, Joanna and many more.

—————

July 7.h.

The fine, warm weather continuing.

We spent part of the morning in the National
Gallery.

We dined with Minnie:—met the Lambton
Loraines, Admiral and Mrs. Gordon, Mr. Sinclair,
Miss Julia Moore, Susan and Joanna Horner and
Edward, besides John Herbert who acted host :—a
pleasant party.

—————

July 8th.

We went to morning Church at Eaton Chapel.

We had the great pleasure of a visit from dear
Lady Muriel and Courtenay Boyle.

—————

July 9th.

We went (Minnie with us) to see the Gibson and
Diploma Galleries at the Royal Academy.

Our dinner party :—Lady Muriel and Courtenay
Boyle, Aberdare and Sarah Bruce, the Leckys, the
Roundells, Lady Alfred Hervey and her daughter,

1883. Mr. Walrond, Sir Brampton Gurdon, Susan Horner, Minnie and Harry.

———

We went (Minnie with us) to see Bull's Orchids in King's Road—magnificent.

We dined with Lady Mary Egerton.

———

Arthur arrived.

We went to the wedding of Arthur Lyell with Miss Florence Chambers, and afterwards to the "breakfast" at the house of the bride's aunt, Mrs. Wills:— met a great many friends.

Called on old Lady Lilford, drank tea and had a long and very pleasant chat with her—Minnie Powys with her.

———

Our dinner party:—Sir Bartle and Lady and Miss Frere, the Leopold Powyses, Lady Octavia Legge, Lady Rayleigh (the Dowager), the Mac-Murdos, Sir Henry and Lady Loch, Alexander Kinglake, the Matthew Arnolds, the Morewoods, Mr. Gambier Parry and General Whitmore.

———

Mr. and Mrs. Wood *(she* was Miss Tollemache), Sir Lambton and Lady Loraine, Cissy Goodlake,

Admiral Spenser, John Herbert and Clement came 1883.
to luncheon.

We visited Lady De Ros, a most interesting old
Lady.   We dined with the Leckys.

July 16th.

Our   dinner   party:— Sarah   Seymour,   Lady
Charlotte Legge, Mr. and Mrs. Cyril Graham, the
Loraines, Mrs. Forbes and and W. Hervey.

July 17th.

Cissy and Emmie staying with us.   Annie Camp-
bell and her brother Colin came to luncheon.   A
visit from Lord Tollemache.   Sarah Seymour, Lina
Bruce, Minnie and John Herbert dined with us.
Fanny   went   to   the   Frere's   garden   party   at
Wimbledon.

July 18th.

Sent off to Spottiswoode the *final* portion of the
MS. of my book,—Fanny and I having together
done the index.

Mr. and Lady Evelyn Upton drank tea with us,
as well as many other friends.

Cissy and Emmy left us.   We dined with the
Aberdares.

July 19th.

We went on a visit to the Charles Hoares at
Minley ; Mr. Church staying with them.

July 20th.

I had a pleasant though solitary walk in the fir
woods.

The John Martineaus arrived, also Colonel and

1883. Mrs. Moncrieff; Mr. and Mrs. Fitzroy and Mrs.
Elliot and others came to dinner.

At Minley.

Violent storms of rain—but we walked through
the gardens with Kate and Mrs. Martineau and
Mrs. Moncrieff. We had a very pleasant little drive
with Kate in the pony-carriage. Lady Muriel and
Courtenay Boyle arrived.

At Minley Manor with dear Kate and Charles
Hoare: we went to morning Church with them.

We parted from our dear friends at Minley, and
came to London in company with Courtenay and
Lady Muriel Boyle and Colonel Moncrieff.

Katharine, Susan, Joanna and Minnie came to
see us in Eaton Place.

Received from Spottiswoode proof of *final* portion
of my book, to end of index.

We came to Lady Grey's, Fairmile House. Mr.
and Mrs. Earle came to dinner—pleasant people.

At Lady Grey's—she kind and interesting as ever. Weather cold, variable and ungenial. Lady Grey took us to call on Matthew Arnold, but they were out: afterwards to Thames Ditton, where we saw Annie Campbell and several of her brothers and sisters.

————

July 26th.

Walked a little on the Common with Lady Grey. In the afternoon she took us to visit Mrs. Earle, and afterwards to Fox Warren—Mrs. Charles Buxtons—a beautiful place, where were a large company of relations.

————

July 27th.

From dear Lady Grey, we went by road to Richmond, a pleasant drive, and spent the afternoon with William and Mimi Bruce:* they both most kind and pleasant. Thence in same way to Eaton Place. Edward dined with us.

————

July 28th.

48, Eaton Place.
Lady Wilmot came to luncheon. Susan, Joanna, Minnie, John Herbert, Leonard Lyell and Susan Zileri dined with us.

————

July 29th.

We went to morning Church at St. Peter's,

* There I first saw my little dog Ruby.

1883. Eaton Square, in Mrs. Storr's pew: the Clements Markhams came in after Church to see us.

_____

Called by appointment on Miss North, and spent a good time very pleasantly in looking over her wonderfully rich collection of beautiful drawings of Cape plants. Minnie dined with us, and went with Fanny and Arthur to the evening exhibition at the Fisheries.

_____

Morning very fine; a storm in the afternoon. Down to Barton, Joanna with us: arrived safe at home, and found all well at our dear home—thank God.

_____

Lady Rayleigh, Mrs. John Paley, Hedley Strutt, and several others came to luncheon and spent most of the afternoon here.

_____

We went to see poor Jarrold, who is dying.
Fanny went with some of the company to see Ickworth.

_____

We all went to morning Church. An excellent sermon from Mr. Harry Jones.

LETTER.

Barton Hall, Bury St. Edmund's,
August 7th, 1883.

My dear Edward,

A very old friend, whom very likely you 1883. may not have seen, for many years, is dead—Miss Bucke. She died last Saturday at Yarmouth, whither she had gone for the benefit of the air. She has been such a sufferer from ill-health of late years (being nearly blind, beside other maladies) that this must probably be looked upon as a happy release for her. She was, I believe, within a few months of the same age as myself. I am sure you remember our playing *bagatelle*, and other games with her at Mildenhall, in those days which now seem so far away.

We had a piece of news this morning, of a very different kind, which it is possible you may not yet have heard:—Mimi Bruce has given birth to a little boy.* I am very glad for her and Willie, and for both families.

We are both well, and have the house nearly full of guests, almost ever since we arrived:—Joanna, Mr. and Mrs. John de Grey, Mr. Heathcote, Clement, Arthur for two days, and now Leonora and her daughters. The men of this party (except Arthur) were drawn hither by the Assizes. Mrs. De Grey, who is but lately married, and was almost a stranger, is very nice, lively and natural.

Ever your affectionate brother,
CHARLES J. F. BUNBURY.

* Fox.

JOURNAL.

1883.   Lady Hoste and Miss West came to luncheon. Lady Bristol and Mr. and Mrs. Seymour afterwards to see our pictures.

———

Dear Joanna went away.   A most beautiful day, very warm.   Drove with Fanny and Leonora in the afternoon, visited the Claughtons at Fornham St. Martin.

———

Another splendid day.   Our dear Arthur Herveys (the Bishop, Lady Arthur and Caroline) arrived.

———

Fanny took a drive with the Bishop and Leonora. Mr. Sinclair arrived.

A very large dinner party; nineteen in all. Bevans, Algernon Bevans, Harrises, Phillipses, Mr. Scudamore.

———

A fine afternoon.

Archdeacon and Mrs. Chapman and a daughter, and Mrs. Wilson and a daughter arrived.   Mr. Beckford Bevan dined with us; also the Claughtons and Mr. Upcher.

The Arthur Herveys went away, also Mr. Sinclair. Mr. and Mrs. James dined with us.

————

August 18th.

A brilliant day, very warm. The Chapmans and the Wilsons went away. Arthur arrived. The harvest going on well.

————

August 19th.

A splendid day, very hot. We went to morning Church, and (together with Arthur) received the Communion.

————

August 20th.

Fanny and Leonora went to Mildenhall, and spent the day there, returning to dinner. I took a short walk with the girls.

————

August 21st.

Again a splendid hot day. Arrangement of grasses in my herbarium, according to Bentham.

————

August 23rd.

A most beautiful day. Dear Leonora and her two nice girls left us.

————

August 24th.

Another glorious day. Mrs. Mills arrived. The wheat harvest finished in this parish, or very nearly so.

August 25th.

Drove with Fanny and Mrs. Mills to Norton, and called on the Henry Hoares.

———

August 26th.

Yesterday I received from Spottiswoode 25 copies of my "Botanical Fragments." I have noted in this journal, under December 13th of last year, the beginning of the printing of it : and on the 9th of June (soon after we arrived in London), I revised the *last* sheet—the *index* excepted. This index proved a very troublesome and tiresome job; my dear excellent wife undertook it for me, and worked at it for an hour or more every day : but I (of course) revised every page of her work, and with the multitude and diversity of other occupations in London, it was very annoying, and consumed a great deal of time. It was not till the 24th of July that I returned (corrected) the last proof sheet of the index. Now that the printer's work is finished, I am very well satisfied with it :—the printing is remarkably correct (indeed all through I have been agreeably surprised with its accuracy), and the type is very good. Now comes the business of distributing the copies, for it is "privately printed."

———

August 27th.

Fanny and Mrs. Mills went to a garden party at the barracks.

We are blessed this year with glorious weather for the harvest, an advantage which we have not

enjoyed for many years. For nearly the last fortnight the weather has been beautiful (with the
exception of the 15th and part of the 16th), and
last week especially was splendid—glorious sunshine
without oppressive heat, unless quite in the middle
of the day. Accordingly the harvest has gone on
quite as we could wish; in this parish the
whole (I believe) of the wheat has been cut and
carried within a fortnight, and it is in excellent
condition. The barley harvest has now begun, and
that crop also promises (I am told) as well as
possible. In fact there seems great reason to hope
that the harvest of this year will be much the
best there has been for seven or eight years.
We ought indeed to be deeply thankful to God
Almighty for such a blessing.

———

September 10th.

Susan Horner said to-day—that according to her
experience—life has in it much more of pleasure
than of pain. I heartily agree with her: my experience and sentiment are entirely in agreement
with hers. Of course the case must be very
different with those (very many I fear), who can
only recal the experiences of extreme poverty or of
life-long disease. For myself, I have to lament the
loss of many dear and highly valued and much
regretted friends: but this, to a great extent, is a
misfortune inseparable from human life—inevitable
to all who live through the usual span of life.
Excluding these losses by death, and the frequent

1883. recurrence of stings of conscience on account of old follies and sometimes of sins, I must say that my memory recals much more of what is agreeable in the past than of the contrary. Unlike Dr. Johnson, I can honestly say that I can remember many days—aye, and weeks and months, which I would gladly—if the choice were offered to me—live over again : and I am very grateful for the power of recollecting them.

_____

September 18th.

As fine a day as possible : we all went in the open carriage to Mildenhall and returned to dinner. Saw the Livingstones and the schools. In the girls' school, the reading and singing was remarkably good.

_____

September 20th.

Mr. Lockwood, the Poor Law Inspector, arrived. He is a very pleasant man.

_____

September 30th.

Received a very interesting and very gratifying letter from George Bentham, concerning my book, of which I very lately sent him a copy. The dear old man (he is now at least 83 years old) is now very infirm, so much so that, as he tells me in this letter, he is no longer able to go to Kew ; this must be a great privation, as he has for many years past been in the regular and constant habit of going thither every week day to work in the herbarium. He was very ill in the spring and early summer, so

that his death was thought to be close at hand : and 1883.
though he has rallied in some degree, I do not
know whether he has been out-of-doors since.
But his mind is in perfect preservation, evidently.

It is very happy that Bentham and Hooker's
great work, "The Plantarum," was completed
before the former was pulled down by this illness,
so that he was able to take an active and important
part in it to the last. The very last family in it—
that of the Grasses, was worked up principally by
Bentham, and it is done in a masterly way. The
care and labour which he bestowed on this very
difficult order of plants, and the ability with which
he treated it, are truly remarkable in a man of his
age.

[During this autumn we had the following visitors
staying in the house :—

Susan and Joanna Horner, Edward, Clement,
Minnie Napier.

Mrs. Mills.

The Louis Mallets.

Mr. and Mrs. Locke (she was Isabel Hervey) and
William Hervey.

General and Mrs. Ives.

Mr.* and Mrs. Locker.

Mrs. Swinton.

And during the same time we had as morning
visitors, or at dinner—

The Duke of Grafton.

Lord Strathnairn (he was Sir Henry Rose).

Lady Hoste, and Mr. Greene, and Dorothy.

---

* Mr. Locker was the author of London Lyrics.

1883.

Mr. Milner Gibson and Gery Cullum.

Mr. and Mrs. John Paley.

Lady Rayleigh.

Mr. and Mrs. Bland and Mr. and Miss Abraham.

F. J. B.]

_____

2nd October.

We (Minnie with us) went in the landau to Mildenhall, found everything in good order at the house.

Received Spottiswoode's receipted bill for the printing of my " Botanical Fragments:" the total amount £72 4s. 6d., considerably less than I had expected. I am glad to have got the business finished, and thankful that what I have done is approved by my friends and by such first-rate judges as Hooker and Bentham.

_____

3rd October.

Mildenhall.

The girls' school, very good reading and singing. The bad weather kept me in doors all day. Read Sir James Stephen in *Nineteenth Century*, on "British India."

_____

October 4th.

At Mildenhall. Shut up by the terrible weather. Read Jessop in *Nineteenth Century* on "Clouds over Arcady."

_____

We returned home.

On the 28th of last month we had the melancholy news of the death of our dear friend Mr. Henry Bowyer.

Our acquaintance with Henry Bowyer dates from between 20 and 30 years ago, beginning when he was appointed an Inspector of Workhouse Schools. He visited us every year at Mildenhall and here, and we became quite intimate. He was a good, conscientious, honourable man, of a sensitive and affectionate disposition, very well read. He was far from leaning to the Romish doctrine which his brother Sir George had espoused.

_____

October 6th.

Louis and Fanny Mallet, with one of their sons, John Herbert, Mrs. Frederick Campbell and her son, and my great-nephew Cecil arrived.

_____

October 8th.

Sent to Nicholl and Manisty an order on Drummond for payment of Lord Sandwich's claim.

=======

LETTER.

Barton Hall, Bury St. Edmund's,
October 11th, 1883.

My dear Katharine,

1883.
I am very glad to hear that you like my book,
—your remarks on it in your last letter to Fanny,
are very interesting to me. I have had (as you
have already heard) very interesting and very gratifying letters from Mr. Bentham and Joseph Hooker
—showing that both of them have already paid
more attention to it than I could have expected,
considering the one is so busy, and the other such
an invalid. Bentham is really a wonderful man ;
his masterly paper on grasses, in the last volume of
the *Linnean*, shows not only astonishing power of
intellect, but astonishing power of work in a man
over eighty years old. How fortunate that his share
of the Genera was finished before his last illness.

I am re-arranging the grasses in my collection
according to his paper. I should be particularly
glad to send a copy of my book to Dr. Asa Gray,
for whom I have a great admiration—if you would
be so good to tell me his address, and how a book
can be sent to America.

I have been much interested by your account of
Kinnordy You will have heard all about us from
Susan and Joanna, as well as before, from Leonora
and her girls : I enjoyed their company very much,
and was very sorry to part with them. The beautiful weather, of which we had so much in September
(especially in the middle part of the month), was a
great enjoyment to me, and I wish you could have

seen the flowers in our long bed opposite the houses 1883.
—such profuse blossoming of Tritonia, Clematis,
Phloxes, single Dahlias, Anemone Japonica, and
most choice of all, the splendid variegated scarlet and
yellow blossoms of Tigridia pavonia. This last I
had not seen in perfection since the time when my
mother cultivated it ; each flower lasted only a few
hours, but there was an almost daily succession of
them for very nearly a month. And with these
flowers there was such a profusion of butterflies as I
have seldom seen here. Peacocks, Scarlet Admi-
rals, Tortoiseshells, and (two or three times),
Painted Ladies.

We were very unlucky in our visit to Mildenhall ;
it rained and blew with fury the whole of the two
days we were there.

Ever since we came back we have had the
house full of company—very agreeable people ; and
I have been particularly glad to have the Louis
Mallets here. I am very fond of *her*, and I consider
Louis one of the wisest as well as most interesting
men I am acquainted with. Presently the guests
for the Bury ball will begin coming, and I suppose
we shall be in a pepetual racket for a month at
least.

I am just finishing " The Life of Lord Lawrence,"
—very interesting indeed, but I think it would bear
a little shortening.

With much love to Rosamond, and as many as
may be within your reach,

Ever your loving brother,

CHARLES J. F. BUNBURY.

JOURNAL.

1883. Had a pleasant walk with Minnie. Lady Rayleigh (senr.) and the John Paleys dined with us. Mrs. Swinton arrived—also Clement.

————

A beautiful day.

Had a pleasant letter from Clements Markham, acknowledging my book.

I wish I had time or memory enough to note down half the interesting things I have heard from Louis Mallet since he has been staying here with us. A few desultory notes I must attempt. He read, while here, Wallace's articles in MacMillan on "The Nationalisation of the Land:" his opinion of them was, that they showed a really surprising ignorance of the very first principles of economical science. He was surprised that a man—whose name stands (deservedly) among the very highest in natural science—should voluntarily undertake to make such a display of his ignorance in another subject.

————

Louis Mallet said that the writer in the *Nineteenth Century* (Keay), whose article on "The Spoliation of India" startled me so much, is not at all trustworthy. Sir James Stephen on the other hand, he says, runs into the opposite extreme, and paints the state of British India in colours much too favourable.

Louis Mallet explained to us that the outcry against the so-called "Ilbert Bill," and the alarm as to its probable working, are much exaggerated; that the number of natives to whom it can give any power is very small, and if it does little good, it can do little or no harm.

The other day, at a fancy ball (at Calcutta, I think) there was one dressed up to represent the "Ilbert Bill," and labelled accordingly. Mr. Ilbert, himself, was asked by a friend what he would do if he met this travestie of himself:—he replied—" I would read him three times and pass him."— (Auth.—Louis Mallet).

Louis Mallet said that the frequent attempts which have been made by our Government to educate and train Indian Princes with a view to elevate their characters and render them more fit for high positions, have been generally unsuccessful: even when brought up as Christians, the influences of the Zenana are too powerful.

———

October 15th.

We have had a very interesting and valuable group of friends and acquaintances staying with us since the 6th of this month:—Louis Mallet, an admirable man: his wife worthy of him, and delightfully cheerful and light-hearted, William Napier

1883. and Montagu and Susan MacMurdo, Mr. and Mrs. Sancroft Holmes, a most agreeable young couple, she very handsome and *accomplished*, — he very intelligent.

===

LETTER.

Barton, Bury St. Edmund's,
October 16th, 1883.

My dear Katharine,

You are quite right about *Bauhinia ;* it was named after the two brothers John and Caspar Bauhin, who both of them were great botanists in their time, and the twin leaves were thought to be symbolical of the brothers.

We still have beautiful weather, and the long bed and the wall behind it are gay with flowers.

It is a long time since we have had such an enjoyable autumn.

Our large party is mostly dispersed ; but we still have the dear Campbell girls, Annie and Griselda ; also Minnie, Emily Napier (the younger), Mrs. Swinton, and the two charming little Seymour boys.

Edward has started for Marseilles, on his way to Barcelona, Majorca, and Algiers.

I hope Susan and Joanna will have fine weather for crossing.

Ever your loving brother,

CHARLES J. F. BUNBURY.

JOURNAL.

I see in the newspapers the death of the famous 1883. *palæo-botanist*, Oswald Heer, of Zurich, whose great work on the "Tertiary Fossil Plants of Switzerland," and that also on the "Fossil Flora of the Arctic Regions," had placed him quite at the head of recent investigators of that branch of science. I was not myself acquainted with him, but Charles Lyell knew him well, corresponded with him, and liked him much.

I received from Mrs. Halliday a most beautiful flowering specimen of Lapageria rosea, which had blossomed in the open air against a south wall, at her home, Glenthorn near Lynton, N. Devon. I never saw finer blossoms of the plant, either for size or colour. With us, the Lapageria requires a greenhouse, and indeed has not thriven well; owing, I suppose, to our not well understanding its management.

___

Dear Annie and Griselda Campbell went away.

The Livingstones, Miss Guinness, Canon Grant, and Mr. Barber arived.

The Harry Joneses and the Jameses dined with us.

___

Sent copies of my book to Mr. Harry Jones, Mr. Sanford, the British Museum Library, and Cambridge University Library.

1883.   Our clerical company went away.

The dear little Seymour boys, Charlie and Bill, went away.   Caroline (*Car.*) MacMurdo arrived.

* * *

October 23rd.

Disappointments as to guests.   A chapter of accidents—railway accident and consequent delays. Arrival at last of Leopold Powys and two daughters, Walrond and one daughter, and one son.   General Lnyedoch Gardiner and two daughters, Mr. Rickman. Some of them went to the ball at Hardwick.

* * *

October 24th.

A very pleasant walk with Theodore Walrond. Dancing in drawing-room in evening.

* * *

October 25th.

A most uncomfortable surprise:—one of our guests, young Newdegate, attacked with scarlet fever,—dismay and disorganization of our party—no Bury Ball for them.

Received a most charming letter from dear Rose Kingsley, to whom I had, some time ago, sent a copy of my book.   She has lately returned home from a fortnight's visit to her sister and brother-in-law at Clovelly, and she writes with most eloquent enthusiasm of her delight in that singular and beautiful place.

* * *

Arthur arrived from Ireland. A very lively dance in the dining-room in the evening.

A beautiful day, quite warm.
The Powyses, the two Miss Gardiners, Mr. Rickman, and Cecil went away.

A beautiful day, but I was not able to go to Church. Read prayers with Fanny and Arthur. Arthur set off at 3 p.m. to return to Ireland.

The Walronds, General Gardiner, and Mr. Lacaita went away. Minnie, Car. MacMurdo, and Harry remaining.
Sent my book to the Athenæum, to the Linnean Society, and to the Indian Office for Mr. Grant Duff.

The weather remarkably fine during the greatest part of last month. Three very bad days near the beginning of it, when we were at Mildenhall, but otherwise a decided predominance of fine weather, and several days more like summer than late autumn. The autumnal colouring of the trees and shrubs uncommonly beautiful, — I think 1 have

1883. scarcely seen it more so. Now most of the exotic trees are bare, but the bright yellow of the elms and the golden brown of the beeches are still very rich.

_ _ _ _

November 16th.

I must copy what Bentham says in a letter to my wife, received this morning.

" When I wrote to Sir Charles to thank him for " his "Botanical Fragments," I had only had time " to go through the first paper ; I have since read " the whole carefully with great interest, and was " particularly pleased with the last paper, on " Foliage. I would beg you to convey to him my " repeated thanks for this valuable book, which will " be permanently deposited with the rest of my " botanical library in the Kew Herbarium."

This from one of the greatest botanists of our age, is truly gratifying.

_ _ _ _

November 27th.

I am very glad and truly thankful to say that we are released at length from the incubus of the _scarlet fever._ Young Newdegate, who came to us on the 23rd of October ill with that complaint, and has ever since been laid up with it, was yesterday pronounced free from infection, and took his departure for St. Leonards, with his nurse and soldier servant, who have been his fellow prisoners, Dr. Macnab having made all the arrangements, and to-day we have heard of his safe arrival. I was very sorry for the poor fellow, who had come hither with the expectation of taking part in a ball, and had been

doomed to such a dreary month of imprisonment ; 1883.
but I cannot help being exceedingly glad that he is
gone, as well as very thankful that no one else has
been infected. All the arrangements for *disinfecting*
—burning the furniture of his sick room, fumigating
with sulphur, carbolic acid, &c., have been carried
on busily since his departure.

_____

November 30th.

Very fine weather lately. For several evenings
past there have been most remarkable sunsets, or
rather indeed for some time after the actual sunset,
an extraordinary glow of bright and beautiful
colours—various shades of red in particular, from
orange to crimson—extending far upward from the
western horizon. Some thought it was the Aurora
Borealis, and it had much of the colour and glow of
what we saw in October, 1870.

LETTER.

Barton Hall, Bury St. Edmund's,
November 30th, 1883.

My dear Katharine,
Very many thanks for your letter, as well
as for Dr. Asa Gray's, which I also owe to you, and
which is indeed very gratifying. To have received
such letters as I have from such men as Bentham
and Hooker and Asa Grey, is indeed very pleasant.
I must honestly say that I am surprised at the
gratifying way in which they have received my
"Fragments," and that it makes me, myself, think
better of my book than I did before. If Dr. Grey
wishes to insert a notice of it in any of the

1883. American scientific periodicals, of course I cannot refuse him, though I am not in general very fond of reviews. I am studying the last part of the Genera Plantarum, in which there is a very great deal to study: but I feel strongly the truth of what you say about the difficulty of finding time in these short days to do what one wants—or near it—especially as I want to spare my eyes, and not to read small print or to write a great deal by candle light. As for microscopic work, I feel that I must henceforth avoid it altogether. Not that there is anything actually the matter with my eyes, but I feel that they are not so strong as they have been.

You have heard from Fanny of the departure of young Newdegate, so long our involuntary guest: and of the work of scrubbing and sulphuring and burning, which has since been going on.

I am very glad that you have good accounts of all your belongings, and that you and Rosamond are well,—that you are busy and well employed need not be said. We have had wonderfully fine weather for some time past, and a succession of most gorgeous sunsets, if they are really only the sunset glow, and not (as some people have supposed) the Aurora Borealis. But the glow appeared to me to be rather in the west than in the north.

I have lately had Car. MacMurdo for my companion in my daily walks, and she is an uncommonly nice girl.

With much love to Rosamond, believe me ever,
Your truly loving brother,
CHARLES J. F. BUNBURY.

JOURNAL.

A second visitation of winter.

We have heard this morning from Minnie, that 1883. our friends the Storrs, living just beside St. Peter's Church, have had a narrow escape from a terrible disaster in a gale, for a chimney of the next house (if I understand rightly), being blown down, fell right through their house, from the roof to the dining room, so that it is a mercy indeed that they or their children were not killed. The house is said to have been quite *wrecked*.

From the 5th to 7th (both inclusive), we had the great pleasure of dear Sarah Seymour's company. With her we had some other pleasant guests : Mr. and Miss Sinclair (agreeable and interesting young people), Mrs. Wilson and her daughters Agnes and Constance (Agnes Wilson is one of my especial favourites) and Guy Campbell; besides several of our neighbours as dinner guests. And since this party left us, we have had the agreeable news that Guy is engaged in marriage to Miss Nina Lehmann, of whom we hear most pleasing accounts from Katharine Lyell and from Mrs. Galton. I hope they will be very happy.

December 18th.

Received an interesting letter from Edward, from Algeria,—begun at Batna, December 9th, and finished at Bona, the 13th.

K K

1883.   Dear Car. MacMurdo left us, to go to nurse her sister Wena.

December 20th.

Fine morning with brisk wind.

A very good and satisfactory letter from Harry about his engagement, and a delightful one from dear Sarah, concerning Miss Wood.

Wrote to Harry.

December 22nd.

Cissy, Emmie, Car. MacMurdo returned to us from London; Lily (Willie Bunbury's wife), with her baby and Harry arrived.

December 23rd.

We went to morning Church with Cissy, Lily, Car.; we received the Communion.

Received a pleasant, friendly letter from Sir Alexander Wood.

December 24th.

Harry went away early.   His brother Willie arrived, also Sarah Craig.

LETTER.

December 26th, 1883.

My dear Susan,

I like to keep up my practice of writing my birthday greeting to you in time for the New Year,

K K 2

and therefore I begin a letter, though I have little 1883. to tell you. For you have, of course, heard from Fanny all about Harry's engagement to Miss Wood, and all about who she is, and her connexions and all the pros. and cons. Naturally, I should have been still more pleased if his choice had fallen on one whom I had known and liked, such as Caroline Hervey, or one of the Sanford girls, or Annie Campbell; but I have for a good while wished, and strongly wished, that Harry would marry, and not only because I am a great advocate for matrimony in general, but because I think it would be especially useful to his character. And though we have never seen Miss Wood, all that we have heard of her—and of her mother also, from several of our friends—is extremely favourable, and her connexions are very good, so I hope and trust that all will go well, and that the union will be a happy one for all concerned. Harry is certainly very much in love, which is a very good thing for him. His brother Willie is now here with his wife, and I have been very glad of the opportunity of becoming really well acquainted with Mrs. Willie. I like her, and think her an acquisition to our family; she is cheerful, intelligent, lively, and at the same time gentle and thoroughly lady-like. I am much pleased also with Willie. Fanny has told you of the rest of our party. Car. MacMurdo is thoroughly charming.

As for books, I have not much to tell you; for though in winter I ought to have much leisure for reading, somehow I have not read much. I

1883. have been latterly reading (but have not yet finished) "The Excursion"—Wordsworth's—of which, strange to say, I had never before read more than one book, and that at least 40 years ago. I find in it a great deal more to interest me than I expected;—much, certainly, that is very fine; but he is very diffuse, often very prosy. I do not think that in this, which is considered his greatest poem, and is certainly his longest, there is anything equal to his lines on Tintern Abbey, or to his Ode on the Intimations of Immortality.

I have just finished reading the Reminiscences of Lord Ronald Gower, a book which was mentioned to us by Katharine, and which we have found much better worth reading than I at first expected; it is not only very pleasant reading, but full of good sense and good feeling, as well as of enthusiasm for the Fine Arts. I have lately re-read Romola; I found it on the whole tedious (at least containing *much skip-able* matter): but the character of Romola herself, is a noble one, and well sustained, and that of the wretched Tito is developed with wonderful skill.

(December 28th).—I have had some very interesting letters from Edward from Algeria; he has been delighted with Algiers itself, and has seen much that is interesting, especially an *oasis* in the actual Sahara Desert, containing 150,000 date palm trees. But I wish we could hear of his being now near home; his last letter was finished at Bona, in Africa, and he says in it that he hopes to eat his Christmas dinner with the Colebrookes, at Cannes:

but we have not heard of his arrival there, and are
beginning to be a little anxious.

Now, dear Susan, I will wind up with wishing
health and happiness and every blessing to you
and all your family party throughout the coming
year, and with hearty love to you all, believe me
ever,

Your loving brother,

CHARLES J. F. BUNBURY.

JOURNAL.

December 30th.

Now at the close of another year, I feel myself
bound, as so often before, to offer my humble
and earnest thanks to Almighty God for the very
many and great blessings which I have been permitted to enjoy, while I deeply feel my unworthiness
of any of them. Above all, I can never be sufficiently grateful for the blessing I enjoy in the
society of such a wife, than whom surely no man
ever had a better,—and for our having been allowed
to live together so long in unbroken peace and
harmony. Next, I am thankful for the good health
which we have both enjoyed during this year, in
which she has completed her 69th year, and I have
nearly completed my 75th. I am not free from the
infirmities of old age: of the "Three Warnings,"
I feel more and more the failure of muscular
strength; but it is what I least regret; I am very
deaf, but not utterly so, and perphaps not more

1883. than I have been for many years. What I most
dread is any failure of my sight ; and I have
felt, this year, some warnings (though slight, I am
thankful to say), to be careful of my eyes.

We may be thankful that, when scarlet fever came
actually into our house, (in October and November)
the infection did not spread.

I sometimes fear that my memory is beginning to
fail ; that it is not as good as it was. I perceive,
certainly, that it is more of an *old man's* memory. I
must be on my guard as to this.

The most interesting event, to us, of this year,
has been the engagement of my nephew Harry
to marry Laura Wood, a daughter of the late
Colonel Wood of Littleton, and niece of Sir Alexander Wood, whom I knew at Trinity. It took
us at first by surprise, for we had never even heard
of the lady till Harry's letter informed us that he
was engaged to her ; and indeed it seems that they
had been acquainted but a very short time before he
proposed and was accepted. We hardly knew at
first what to think of the matter ; but since then we
have heard of her from many of our best friends
(some who have seen her lately, and others who
have been long acquainted with the family), and all
concur in high praise of her and of her mother also.
I trust therefore that we shall find her one whom we
can love, and that with God's blessing, this marriage
will prove a source of comfort and happiness to
Harry, to us, and to all connected with him. I am
sure that nothing can do so much good to his
character as a well-assorted marriage. Laura

Wood's brother is married to the only daughter of Lord Tollemache,—a very satisfactory connexion. I have a great respect for Lord Tollemache, and I like his wife very much.

My nephew William has been with us latterly for some little time, with his wife and child. He is a very fine young man, very pleasing, earnest in his profession, and (if I am not much mistaken), of very considerable intellectual power; much superior to the common run of young men of the day. His wife (formerly Miss Ramsay), whom he married in India, was till now almost a stranger to us, though they have been married some time, and have a fine baby. I like her very much. She is well educated, intelligent, kind, sympathizing, gay and lively, at the same time thoroughly lady-like and affectionate.

We are happy, this year, in having lost very few friends ;—hardly any, I think, except Lord Talbot de Malahide, Mrs. Rickards, and Mr. Matthews (my old tutor); and of these the second was very old, and so infirm both in mind and body, that death could not be unwelcome to her ; and of the third, I had seen very little for many years past. Lord Talbot is certainly a loss ;—I cannot conceive how I came to write this paragraph, forgetting the loss of two such old friends as Admiral Horton and Henry Bowyer, both of whom I have commemorated in this Journal ; and who neither of them deserve to be forgotten.

I think I have nowhere mentioned that we have since the spring, lost our dear Mrs. Storrs from this

1883. neighbourhood—not by any means (I believe and trust), from our friendship and affection—but her husband having been appointed the incumbent of St. Peter's, Eaton Square, they are now settled there, and have deserted Bury. I trust we shall see them (or at least *her*), often, when we are in London, as our house in Eaton Place is within a few yards of theirs.

Milton Keynes UK
Ingram Content Group UK Ltd.
UKHW032320161024
449665UK00001B/27